POULTRY FANCIER

ARTIFICIAL
INCUBATION
&
REARING

For advice on Natural Incubation read:
*Natural Incubation
&
Rearing*

**Also by
Dr J Batty**

ARTIFICIAL INCUBATION & REARING

Joseph Batty
Past President: Old English Game Club

Beech Publishing House
The Bindery
Sawmill Buildings
Stedham
Midhurst
West Sussex GU29 0NY

© J Batty 1994

This book is copyright and may not be reproduced or copied in any way without the express permission of the publishers in writing.

First Edition 1994
ISBN 1-85736-051-6

Beech Publishing House
The Bindery
Sawmill Buildings
Stedham
Midhurst
West Sussex GU29 0NY

CONTENTS

1. Introduction: Meaning of Incubation 1
2. The Egg & Incubation 17
3. Formation of the Egg 29
4. The Reproduction Process 35
5. History of Incubators 43
6. Justifying the Expenditure 67
7. The Incubator Room/Building 77
8. Small-Scale Incubation 95
9. Large Scale Incubation 117
10 Incubator Problems 129
11. Final Developments & Hatching 153
12 Rearing 164
    Index 181

DEDICATION
Dedicated to the memory
of
Charles E. Hearson
who made scientific incubation
possible by his invention in 1881
of his **thermostatic capsule.**

PREFACE

Incubation is a fascinating subject which has received the attention of man for hundreds of years.The Egyptians and Chinese reached levels of great skill and efficiency, yet they had no sophisticated devices.

Charles Hearson invented the regulating capsule towards the end of the last century and from that point great strides were made and the ordinary farmer could achieve very good results.

Then came the age of the mammoth (cabinet) incubator where thousands of eggs could be set in one machine with the expectation that more than 75% would be hatched. It had the impact of turning poultry farming from a cottage industry into a massive trade which transformed diets and habits of eating.

This book is concerned with the modern methods of incubation at all levels. A middle-of-the-road approach has been adopted. Whilst acknowledging that scientific facts must be stated an effort has been made to keep the language as simple as possible, thus allowing the poultry keeper to use the book in his work. Also the needs of the fancier have been considered as well as the large scale breeder.

In adding this title to the Poultry Fanciers' Library we hope once more to be able to serve those who are interested in our domesticated birds who give so much pleasure and provide valuable food.

My grateful thanks to those who assisted with the work and the companies who gave information and photographs of incubators, where possible being acknowledged in the text.

J Batty

1
INTRODUCTION: MEANING OF INCUBATION

A Basic Definition

Incubation is the process whereby eggs are subjected to the *appropriate level of heat, moisture and ventilation,* and turned daily, so that after a specified period, dependent upon the species involved, an embryo is developed and this pecks its way out of its shell. Generally speaking all birds can incubate their eggs and do so. There are exceptions: the Mallee bird of Australia which incubates by burying its eggs in a mound and certain domesticated poultry, such as the Mediterranean breeds (**Italy**: Leghorns and Ancona and **Spanish**: Minorcas and Andalusians) and other light, laying breeds.

The process may be carried out by the bird itself which has many advantages*, but has the major disadvantages that in the case of any domesticated bird there is loss of production whilst the bird is sitting and coming back into laying condition (as long as 8 weeks or more for poultry) and there is a limit to the number of eggs which can be hatched at any one time.

---

*See *Natural Incubation & Rearing* , J Batty

The alternative to **Natural Incubation** is **Artificial Incubation** which uses a machine, often electronically controlled, whereby large quantities of eggs can be hatched ranging from as few as 50 poultry eggs up to many thousands at a time.

In attempting to copy nature it is necessary to consider:

1. Suitable incubator for the task, getting an incubator slightly larger than likely requirements, but not excessively large or it will have to run with many empty spaces which would be quite uneconomic.
2. Selection of a hatching room which does not have too much variation in temperature and can be ventilated. Avoidance of direct sunlight into the room or building is very important because this is the most likely reason for wide variations in the conditions. A well insulated building should be the aim with ample storage for eggs and other requirements.
3. The training needed for successful hatching of eggs. Whilst the incubator manufacturers instructions are vital some experience and training is essential because, if there are difficulties, the operator must be able to deal with them.

An incubator involves a capital outlay so for a poultry farmer its purchase would have to be justified in terms of output and profit. The fancier buys one because he wishes to pursue his hobby.

Advantages of Artificial Incubation

Provided the stock person wishes to hatch more than a few sittings of eggs each year it will be valuable to consider obtaining an incubator. Whilst for the fancier who wishes to breed just a few chicks from his stock the natural method will be quite adequate, the poultry farmer and fancier with many birds will find an incubator quite invaluable. The advantages claimed for an incubator are:

1. Larger numbers of eggs can be hatched.
2. These can be hatched at any time without waiting for a broody hen.
3. The hatching can be timed to the best advantage for such things as:
(a) Show birds – the earlier the better for summer shows.
(b) Winter layers which are best hatched in March or April.
(c) Table birds – these are best hatched early although the broiler industry can produce table birds very quickly at any time. However, with the swing back to all things natural the hatching of chicks for rearing outside may call for selection of early dates.
4. Because of the larger numbers hatched the incubator is more profitable.
5. The modern incubator in the hands of trained staff should be less trouble than using many broody hens.
6. Chicks can be reared artificially and more economically.

7. There is no danger of the chicks getting damaged by the hen taking the chicks into long grass or other danger, or by contaminating them with lice or other vermin. Nor will the eggs get broken by a clumsy hen or one which eats the eggs, which can occur when she is not a good broody.

ESSENTIALS FOR INCUBATION

As noted there are a number of essential requirements for incubation, which are as follows:

1. HEAT

The appropriate heat for the particular egg should be known and then the incubator run at that temperature. The aim should be to have the balance of the many variables set so that hatching is maximized. For poultry this will be in the region of 37°C (c 100° F). In fact, when measured just above the eggs it is usual to have a temperature of 39.4oC (103oF). When speaking of desirable temperature there must be an understanding on which part of the egg is being discussed because the ideal will be different for the centre, on the shell and within a prescribed area close to the egg.

The temperature outside the incubator should be considered and if high the temperature in the incubator should be set lower; conversely, if the room tem-

Artificial Incubation 5

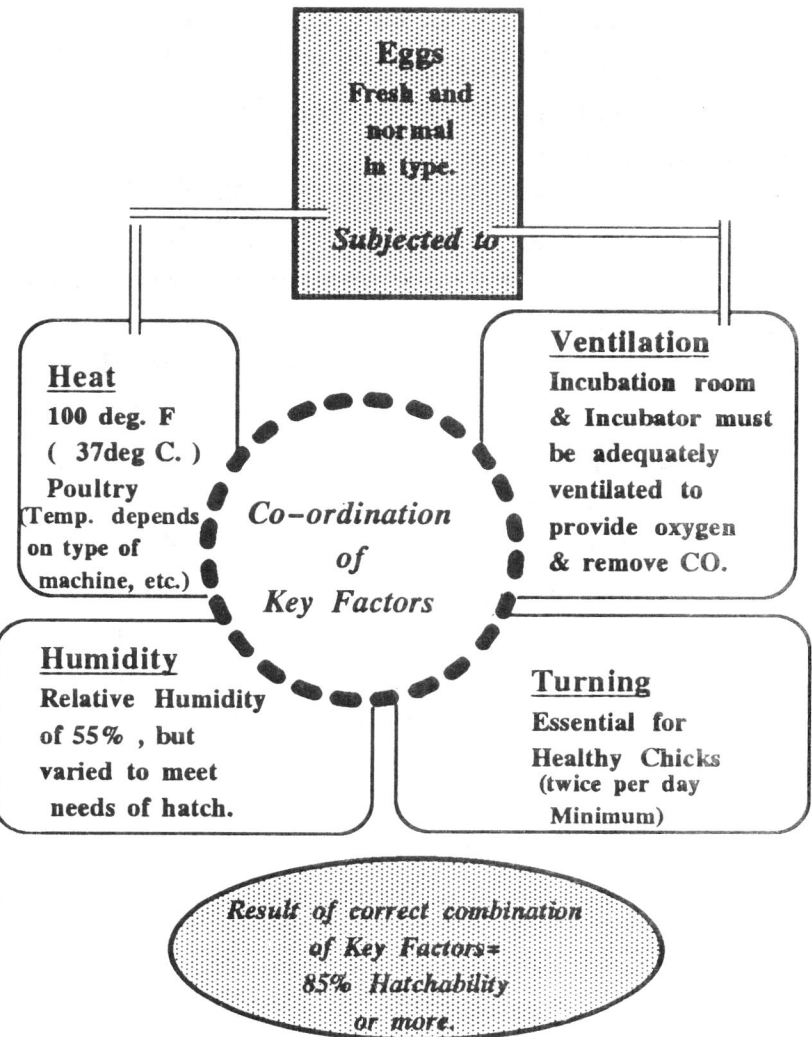

Figure 1.1 Key Factors for Successful Hatching

perature is low, the incubator must be run at a higher level. Generally the temperature in the room should be at a steady level, not exceeding 70°F (21.2°C), but preferably in the scale 55 to 65°F.

A steady temperature is desirable around the optimum level just suggested. Otherwise, the egg temperature will fluctuate with dire consequences. The author has seen many fanciers lose hatches by extreme variations occuring, usually by the temperature dropping dramatically during the night or it rising around noon when the sun shines on to the incubator or through a window in the incubator room.

If too much heat is put into the incubator the air space in the egg will enlarge too rapidly and the embryo will suffer, but if too low the growth will be retarded. Usually, a slightly reducing temperature is to be preferred as the incubation proceeds (say, 0.50° F each week*, but this may be varied slightly to suit the machine) because this has been found to produce healthy chicks. It will be appreciated that as the embryos develop they produce heat within the incubator and just before hatching this is considerable. For this reason, great care must be taken to ensure that the temperature

---

* Variations must be watched very carefully for an increase in temperature of above 1 deg. F (0.50 C) for a long period may spoil the hatch.

does not rise very much and certainly not for long periods; turbulent variations should also be avoided.

2. VENTILATION

Like all other life form there must be the means to breathe and this is provided by the air circulating in the incubator. The circulation of air also removes gases and smells which might develop in the egg and incubator. Thus ventilation supplies essential oxygen and removes the unwanted carbon dioxide, but should be done without causing draughts and removing too much moisture. The room should be well ventillated and quite comfortable to work in.

It has been estimated that a chick consumes a daily rate of grammes of oxygen as follows:

Day	Amount
7	0.071
14	0.714
21	1.714

This is the equivalent of 45.4 cubic feet of oxygen at 21 days. At this stage there is an expulsion of carbon dioxide of 23.0 cubic feet. For a large incubator room it will be essential to have induced ventilation by cowl on the roof, which must be sealed and insulated if this method is to be successful. A fan will also ventilate a room, but this does have running costs. Open windows are not satisfactory because they let heat and insects in .

A general guide is that the air should be changed 8 times per day. In practice, very good results can still be obtained by using sound common sense and observing the conditions which prevail. A minimum and maximum temperature can be recorded by having a thermometer to cover the range; these are readily available from a garden centre.

The smells which occur in the incubator room also give a guide to conditions prevailing. It is natural for some smell to be present near and during hatching, but if this is unpleasant the air is probably stale and there may even be 'gas bombs' sitting in the incubator, just waiting to be jarred, when an explosion will occur. This is why doubtful eggs ahould be taken from the incubator when they are 'candled'. If a rotten egg does explode it means that the eggs affected should be washed very carefully in warm water and the incubator fumigated immediately the incubator is free. Fortunately, at 21 days, an egg will not usually have reached the explosion stage so the occurrence is not common.

At the later stages of incubation more rather than less air should be provided, but if eggs of different ages are being incubated there may be difficulty in adjusting to give more because this would affect the eggs in the early stages. Accordingly, optimum conditions should be the aim.

3. HUMIDITY

Moisture is absolutely vital for the production of life and without it the embryo is not likely to develop. This aspect is referred to as the necessary **humidity level**. If there is a low humidity the membranes will dry up and the chick will fail to grow and even if it manages to live to the time of hatching there will be difficulty in getting itself out of the shell. It will in any case be late in hatching.

On the other hand, if the moisture is excessive, especially after being deprived of moisture early on, the egg will be too wet and the chick will be virtually drowned in the excess.

The amount of moisture required depends upon a variety of factors:

(a) **Type of machine (some regulate the moisture completely).**

(b) **Room in which the machine is situated and the overall atmosphere prevailing.**

For example, a room in a cellar will be damper than on the top of a hill and this will affect the relative humidity.

Incubation has a shrinking effect on the air space of the egg. Accordingly, the correct level of humidity can be measured by visual inspection of the egg at regular intervals and/ or the weighing of the egg to see what loss has occurred. The problem with weighing is the time and cost which are worthwhile for valuable eggs, but too formidable a task for thousands of eggs. However, this process can be carried out satisfactorily by taking a number of eggs on a statistical sampling basis. The results of a few can then be applied to the bulk.

Generally from 10% to around 15% is the normal loss, but the exact figure will be affected by the species involved; the ideal is regarded as 12%. The loss itself does not occur at an even rate and neither is the moisture requirement. In the early stages there is very little loss and little moisture is required. **Towards the end of the incubation period the humidity is raised until it is very high and this helps the chicks in their efforts to free themselves from the egg.**

The measurement of the relative humidity is achieved by using special instruments. **With electronic controls the measurement can be exact**, but generally the vital facts are provided by the following in many incubators:

(a) **Wet bulb** which shows a reading in degrees F or C. and this is compared with the dry thermometer and a percentage is calculated.

(b) **Hair Hygrometer** which measures the moisture in the air which records the relative humidity direct.

Wet Bulb & Dry Bulb

The wet reading is compared with the temperature and a percentage is calculated. At 100 degrees F, when the air is holding the **maximum amount of moisture** the RH is said to be 100%, which is possibly the level required for hatching. On the other hand, if the wet bulb is 85deg. F, this is the equivalent of a 55 % reading on the Relative Humidity scale. This is the *average* recommended level for normal eggs.

If eggs have abnormal shells, too thin and porous, or very thick, such as Guinea Fowl eggs, it will be necessary to adjust the Wet Bulb reading. For thin shells the reading should be around 87 thus achieving the 12% ideal loss and, in the case of thick shells the reading should be lower, say, 83. There will be difficulty if the eggs are of a mixed type and, if possible, only good quality normal-sized eggs of regular shapes should be used.

For cabinet incubators, measuring in centigrade, the recommended readings are Wet Bulb 31 and the Dry Bulb 37.7. When separate hatchers are used different rules will prevail. In all cases the manufacturers instructions are to be followed.

If the air is too dry the egg will evaporate at an increased rate and this will cause cooling to take place.

Hair Hygrometer

The hygrometer has moving parts which operate when the moisture shrinks the hair and turns the mechanism. This instrument is not popular because the hair becomes dirty with dust and particles. Moreover, its accuracy is not usually very high so cannot be relied upon. It is better to stick to the wet bulb/ dry bulb method or take on the more accurate electronic devices.

Electronic Instruments

These will be incorporated into the newly designed machines and the general principles still apply.

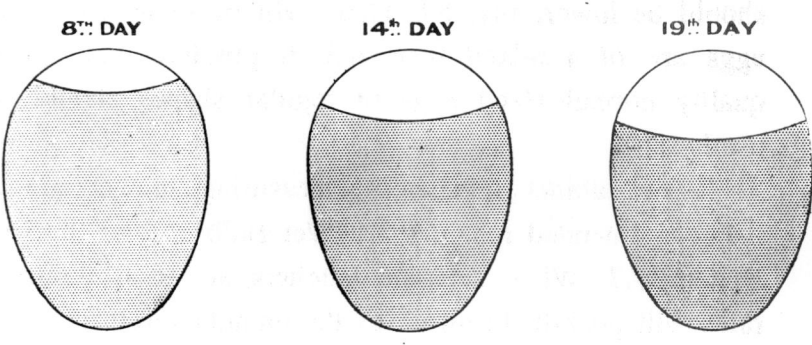

Figure 1.2 The egg and air space

4. TURNING OF THE EGGS

Eggs must be turned each day and as frequently as possible. The broody hen turns her eggs approximately twice per hour. In a small incubator the turning should be at least twice per day; the larger, automatic machine will turn the eggs more frequently, as much as once per hour.

Without the regular turning the embryo will not hatch. The process is vital at the early stages and should be continued up to the 18 or 19th day; no turning should be attempted after that time.

Turning has its affect upon the following:

(a) **Avoids membranes sticking together.**

(b) **Prevents the germinal disc adhering to the inner shell** or the membranes.

(c) Allows the embryo to correct itself when it is positioned wrongly in the egg.

It is estimated that 25% of eggs laid are not in the exactly correct position, with the head to develop pointing to the air space and regular turning appears to correct this fault.

(d) **Brings each side into contact with the heat when the incubator supplies heat from the top.**

This does not apply to cabinet machines because they provide heat which permeates throughout the incubator.

The Aim

In the ideal world we would look to achieving a 100 per cent hatch; this is often achieved with wild birds and the broody hen who lays her eggs in a nest and then proceeds to sit on them. However, with artificial incubation there tend to be more problems and, whilst a 95 per cent hatch is feasible, over a period it will usually be found that an achievement of something *approaching* 90 per cent will be the norm; some farms achieve around 85% and this is regarded as satisfactory, but a higher rate should be the aim.

Records would be kept of each batch of eggs so that regular improvements can be made. As will be shown later, there are many factors which can affect the hatchability of eggs, *many outside the control of the incubator and its operator*. The health of the birds, food given, and the accommodation, are all matters which determine the success rate.

Any of the factors covered above can cause a poor hatch. However, the lack of ventilation is a frequent reason for a high mortality rate at the *later stages* of incubation, usually as a result of carbon dioxide poisoning. At twelve days the chick is in a 'live condition' and breathing the outside air and therefore susceptible to poor ventilation conditions.

Artificial Incubation 15

HATCHING RECORD			Year: 19....			
Breed: OLD ENGLISH GAME (B)						
Hen	Date	Set*	Due	Hatch*	%	Notes
P2	5/4/..	50	26/4	40	80	HATCHED 19TH DAY

Figure 1.3 A Basic Hatching Record
More sophisticated records may be designed for specific needs
*Number incubated and actual number hatched +success percentage vital

Different Temperatures

The references to various levels of temperature throughout this book may confuse. It will be seen later that different types of incubator are run at different temperatures.

A hot water, still air machine is run at 100° F (37.7°C), a hot-air model at 103 deg (39.4 deg C), some of the large table models (now very little used) at 105 deg, and cabinet machines at 100 or 99.50 deg F.; while the hen hatches at 102 deg.

This wide variation is somewhat puzzling to the beginner until it is remembered that the temperature in the centre of the egg is the same in all cases, i.e. 100 deg F (37.7° C.), and it is the different methods of applying heat that cause the variations in the different types of machine.

2
THE EGG
& INCUBATION

Importance of The Egg

The egg serves two basic functions in incubation:
1. Contains the 'germ' from which the chick develops.
2. Provides the food and water on which the embryo develops and on which the chick is given sustenance for 24 to 48 hours after it is hatched.

The albumen is absorbed by the embryo around the sixteenth day. Just after that the amniotic fluid (*Amnio* = sac covering embryo) begins to disappear and on the nineteenth day the yolk starts to be absorbed into the body via the navel. This is a crucial stage and if the incubator is too hot, which starts premature hatching, the yolk will not have been drawn into the body and the blood circulation system will not be in its proper place and the chick will most likely die. This also applies when an attempt is made to remove the chick from the shell before the yolk has been absorbed, thus causing bleeding and sending the chick into a shocked state and, as a result, it generally dies.

Shell

The shell of the egg appears to be smooth and quite airtight, whereas it has many pores (estimated at 6,000 to 8,000 pores in a typical hen egg), with the greater density at the broad end.

The colour is of no significance to the hatching process, except the ability to produce a specific colour egg is passed on. When the calcium carbonate is coated on the egg in the 'passageway' of the hen (known as the uterus) it also receives the colouring of the shell.

This shell is divided into three layers:

1. Cuticle – the outer fine coating.
2. Palisade or sponge layer which is about two thirds the thickness of the shell.
3. Mammillary layer or inner membrane.

The shell is so constructed that it is very strong on the outside, thus allowing a hen weighing 3 kilos to press down on it without being broken. In reality a hen egg of 60 grammes will withstand more than this weight the breaking strength being 4.1 kilos. Interestingly, it is not only the size which determines the breaking strength; for example, a Guinea Fowl egg of around 40 grammes has a breaking strength of 5 kilos and it also has a thicker shell.

Fortunately for the chick, breaking out is not as difficult as breaking in. Moreover, by the time the incu-

Artificial Incubation

bation has taken place the egg shell has become more brittle and fragile. Next time you take the egg shells from the incubator crumble one between your figures and note the difference in the shell between that egg and one that is newly laid.

Creation of a sound, healthy shell is of vital importance in the incubation process. Accordingly, the birds must be given an adequate diet, including a fairly high level of protein, vitamins, amino acids, fats and natural, fresh foods such as greens, apples and foraging material such as leaves and grass clippings. When purchasing pre-mixed pellets there is a guarantee that all the necessary ingredients will be present. This can be augmented by mixed corn and greenstuff, but wheat by itself, containing around 10 per cent protein is not really adequate for breeding, although quite good results can be obtained if birds are on free range and there are other foods available. If there is a thriving insect life and plenty of short, healthy grass, this can make up the deficiency. Game fanciers feed corn on a regular basis because it is believed that this leads to the tight hard feathers required for top class birds, and they have no difficulty in breeding first rate stock. However, numbers hatched tend to be small and the concentration is on physical characteristics rather than on egg laying.

Grit must be provided at all times. If birds are on free range they will find a great deal, especially the small flints used for grinding the food in the gizzard. The soluble grit such as limestone and oyster shell should be available at all times in a grit hopper which is topped up on a regular basis. If birds start to lay soft-shelled eggs this may be a sign of lack of grit. Sometimes it is an indication that the bird is too fat or has some health problem which is preventing the formation of the shell. The problem should be tackled by giving extra grit in the food in powder form and when obesity is suspected more space for exercise and a controlled diet will be essential.

Soft-shelled eggs or those which are very porous should not be used for hatching. If an egg is badly 'pocked' or has an irregular surface without the necessary bloom of the healthy egg, do not attempt to hatch it. Such attempts usually lead to failure and if successful the chick is not likely to be healthy.

On this question of hatchable eggs it should be noted that irregular shapes are best avoided and so are abnormally large or very small eggs. For a typical large fowl the ideal size is 50 to 60 grammes. Remember that egg size and shape are factors that are passed on to the next generation so care in selection is important.

Artificial Incubation

Table of Egg Sizes

Breed/Species	Grammes
Domestic Fowl	58 – 68
Domestic Duck	80
Domestic Turkey	85
Guinea Fowl	40 – 50
Bantams	30 – 48
Geese	200
Partridge	18
Pheasant	32
Peafowl	90
Ostrich	1,400

Note: As will be shown later, the incubation periods differ from one species to another.

Figure 2.1 Relative Sizes of Eggs

Normal Eggs

The aim should be to hatch from eggs which are a model of what is required for future generations. Therefore, in hatching, the following should be considered:

1. Parent Stock*

The parents should have the qualities desired on both sides. If the prime objective is for first-class laying birds then eggs should be selected from top quality birds. The hens can be trap-nested and the cock should be from the same strain, full of vigour and vitality.

When dealing with show birds select birds with those qualities that are likely to produce the features or 'points' which come nearest to the *standards*.

The desired level of hatchability will only be produced from stock which has the necessary qualities and is fed on suitable food.

2. Shape

The egg of most birds is **ovoid** in shape; that is, it is rounded, but is longer than the width and has a blunt end where the air space is to be found and the 'pointed end' which should also be rounded. The normal shape may be seen from **Fig. 2.2 Shape of Eggs**, alongside other eggs which are best not hatched. Because the heat is measured from a specific point (eg, just above the eggs) the eggs being incubated are best if around the same size. Abnormal shapes often indicate there is disease or some other physical problem.

---

* Readers wishing to pursue breeding and related aspects are advised to study *Practical Poultry Keeping*, J Batty.

Artificial Incubation 23

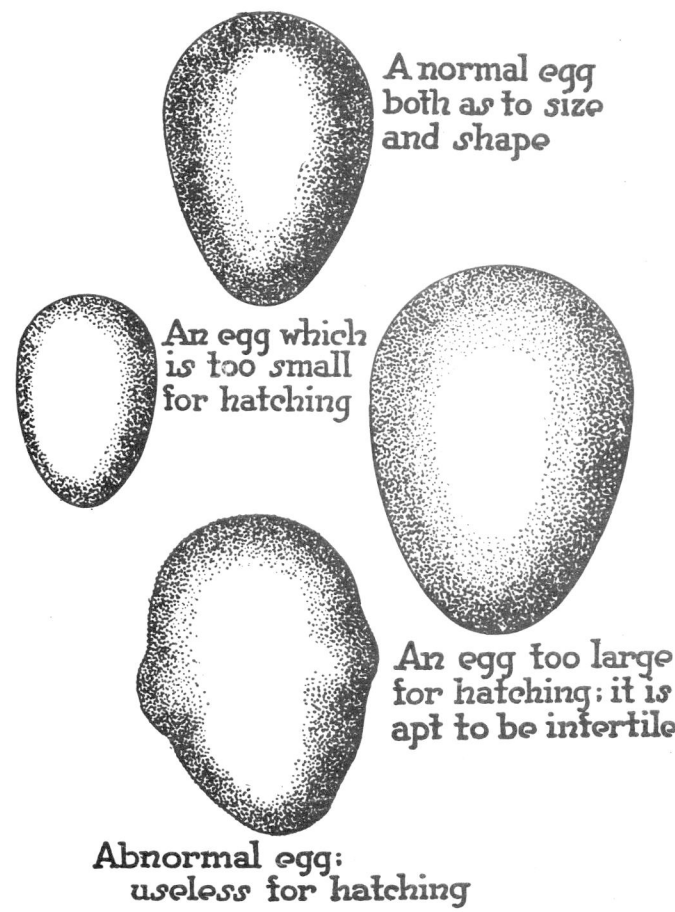

Figure 2.2 Shape of Eggs
Only normal shaped eggs of sound quality with excellent shell quality should be incubated.

3. Internal Quality

Eggs should be free from any substances which are abnormal in type. These would include blood and meat spots, double yolks, an unusual colour or other abnormality. The essentials of acceptable eggs are:

(a) **Egg white** – should be clear and free from mould or other defects.

(b) **Yolk** – a sound, normal size, yellow yolk should be in the normal position, not stuck to the shell inner membrane or sunken or mispositioned in any way.

(c) **Size of air space** – a fresh egg has a small space at the blunt end of the egg which grows as the egg ages. Any indication of too large a size must call for a query to be raised because the egg may be old or has been stored incorrectly. (see Fig. 2.3 The Air Space and its Enlargement)

4. Shell Appearance.

The shell should have a 'bloom' to it and be free from hair line cracks and excessive mottling or pitting. Excessively dirty eggs or those which have not been washed immediately and are soiled should be discarded. Any bacteria on a shell can invade a dirty egg very quickly so there is little point in attempting to incubate such eggs which might contaminate the others in the trays.

The defective eggs can be detected by candling and for large numbers the machine allows the egg to be rotated over the light, thus enabling faults within the egg to be detected. Hair cracks and other faults in the shell

Artificial Incubation 25

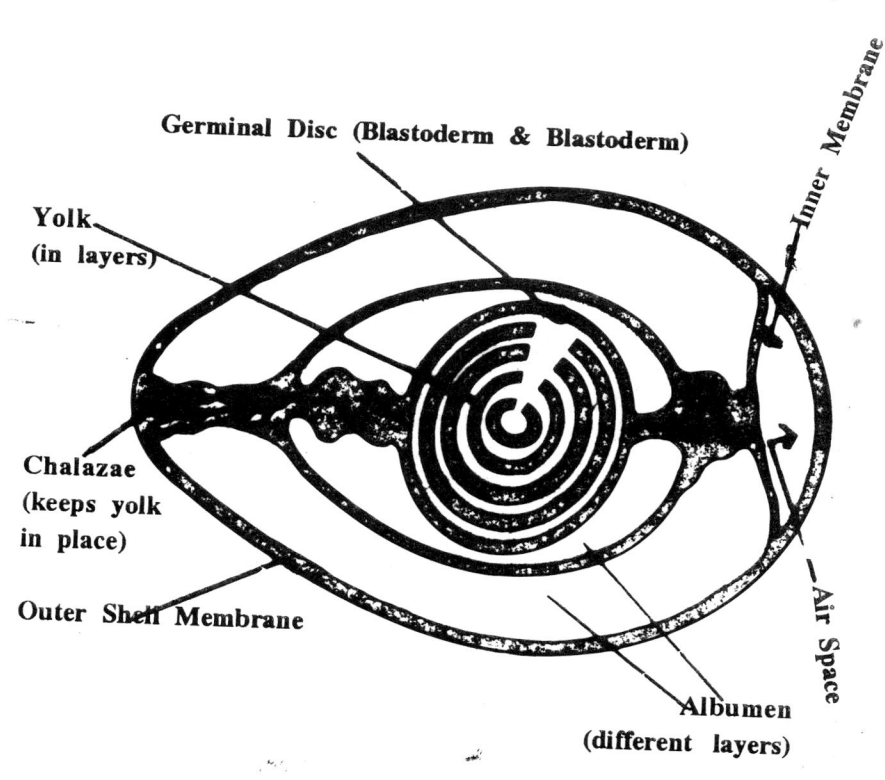

Figure 2.3 Main Parts of the Egg
These change in the process of incubation.

can also be seen in that way.

Generally the young laying birds will have the best looking eggs, but for various reasons the selection of hatching eggs should be from mature birds, in their second season. Size of egg is better, there is a more complete record of ability to lay and chicks will tend to be stronger. However, as a bird gets older, the danger of faulty eggs increases. The shells tend to be thinner, they may be more porous and blood and meat spots may be present because these appear in birds which have been subjected to more stress and strain.

Getting the Hatchable Egg

Many factors will affect the number of eggs which can be incubated and *prima facie* offer up to a 90 per cent chance of hatching. At this stage it is useful to summarize these requirements:

1. **Adequate food, water and grit.**
2. **Sound healthy stock** with a cockerel or very virile cock matched to a reasonable number of hens. For the fancier this means a hen and cockerel or a trio. For layers around 12 birds per male may be adequate and a smaller number for heavier breeds, but the method of management will have a bearing on the numbers. The guide is to get a balance which will produce a high rate of fertility so tests should be carried out as soon as eggs are available.

3. Nest Boxes

These should be cleaned regularly with the addition of new nesting material such as straw or shavings and located in suitable places which afford privacy.

4. Egg Collection

Collect eggs on a regular basis (four times a day) and ensure these are clean. If necessary clean them by a dry method which means using steel wool or sandpaper or, for large scale cleaning, machines are available.

Washing in liquid is also a possibility, but this requires great care including the use of water at a temperature of not more than 90 degrees F. because a greater heat will damage the eggs, but much lower will not wash them properly. Some experts advocate a higher temperature simply because this cleans better, but there is no proof that high temperatures will be any better than the 90 deg. quoted.* The washing should not be longer than 5 minutes and regular changing of the fluid is essential. Dirty sterilizing fluid would certainly defeat the purpose of the washing and may even multiply the germs.

5. Watch the Birds

An alertness in management is essential so that nothing which affects the laying and hatchability is allowed. Running out of food or water, cocks fighting, hens being bullied, appearance of rats or other vermin and other occurences which would reduce efficiency must be tackled.

---

*In *The Incubation Book*, Dr A F Anderson-Brown the author gives an example where duck eggs are washed with farm detergent and hypochlorite for 3 minutes at temperatures of 140 deg. F., but this seems to be exceptional for duck eggs only.

28 — Artificial Incubation

Broodiness is undesirable (unless a broody is needed) so steps should be taken to remove all broodies from the nest and place them in a different environment.

5. Free Range or Semi-intensive Conditions.

Strong healthy chicks come from natural conditions and therefore if at all possible the breeding flocks should be on free range. Smaller numbers may be in pens, but letting them out each day and throwing in greenstuff and other essentials will encourage exercise and healthy birds.

Figure 2.4 A Typical Egg Washing Machine
The *Rotomaid* – larger machines are available

3
FORMATION OF THE EGG

Creation of the Egg

The egg is the foundation stone on which reproduction rests and therefore an understanding on how it is formed and factors relating to this aspect are vital to the understanding of poultry or other bird management and successful incubation. The ability to lay is present from birth a factor inherited from the parents. Over many generations of selection and sound management the domestic fowl has progressed from the Jungle Fowl, cabable of laying up to say 30 eggs, to some strains that will lay upwards of 300 eggs per annum.

Food must be adequate in terms of quantity and quality to sustain the egg laying potential and a hen not receiving adequate nutrition may not produce hatchable eggs. In fact, without a proper diet the hen may fail in her role as a mother and her health will be endangered.

The Process

The formation of the egg starts from the ovary which is a cluster of eggs, which start as very tiny specks and progressively, develop into full size eggs. There are

separate though closely related operations:
1. Yolk development in the Ovary.
2. Albumen (white) plus two membranes and the shell, which are formed in the oviduct, which is the long 'tube' extending from the ovary to the cloaca from which the egg emerges.

THE OVARY

The ovary is the part of the body located near the left kidney, suspended in the abnominal cavity, which contains all the embryonic eggs. It has been likened to a bunch of grapes, each baby yolk (ova) being on a stalk attached to the main stem.

There are a number of features which should be noted:

1. Both ovary and oviduct are provided in duplicate, but only the left develops.
2. Size - not standard in any way because it varies according to the egg potential.
In the fowl the number of oocytes (egg cells) varies from around 600 up to more than 3,600 and the largest number of eggs laid in a domestic hen has been recorded as around 1,500 so not all potential eggs are produced.
In the active state the ovary increases substantially in size.
3. Ovary colour is a greyish white which turns to yellow when laying commences.

Artificial Incubation

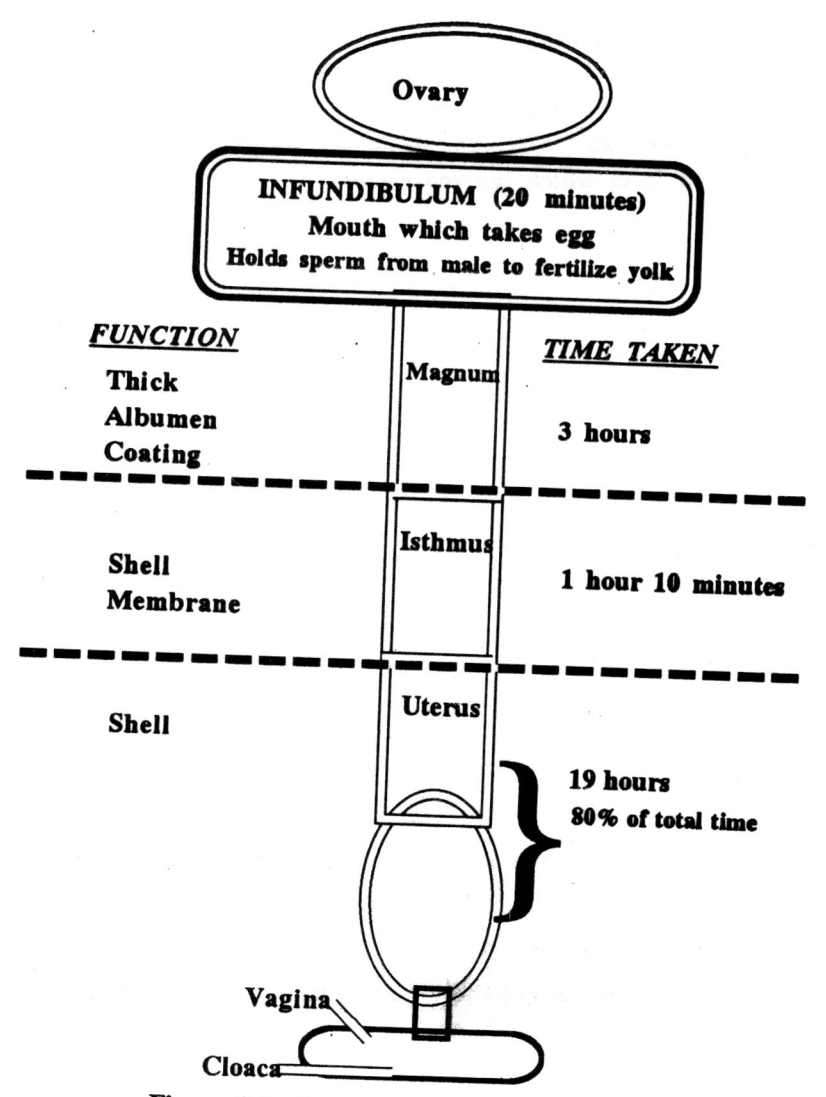

Ovary

INFUNDIBULUM (20 minutes)
Mouth which takes egg
Holds sperm from male to fertilize yolk

FUNCTION		TIME TAKEN
Thick Albumen Coating	Magnum	3 hours
Shell Membrane	Isthmus	1 hour 10 minutes
Shell	Uterus	19 hours 80% of total time

Vagina
Cloaca

Figure 3.1 Stages in Production of Egg
Not to scale; See Fig. 3.2

4. The yolk develops in the ovary over a period of time and at around 6 months the domestic fowl is mature and begins to lay.
The hen will lay from around the age mentioned and requires no sexual or other stimulation although the weather and light will affect the output.
5. Each yolk proceeds to the 'pocket' of the Oviduct and begins its way to the point of lay.
This is done by the yolk rupturing (breaking out of) the follicle (pouch) in which it grows. After about half an hour the next yolk then takes its turn for depositing in the oviduct and is therefore a continuous process, the best layers producing an egg each day As shown below each developing egg takes a considerable period to go through the oviduct to emerge as the complete egg.

THE OVIDUCT

The oviduct is the second part of the process involved in the formation of the egg. Once the yolk has been formed and the follicle ruptured, it goes into the **ovarian pocket,** which is a cavity near the ovary. After that the ovum (yolk) is taken by the **infundibulum** which is the entrance of the oviduct. This process is known as *ovulation.*

There are various stages, each taking a period of time for a function to be carried out. These are the addition of the white of the egg, the membranes, the shell

Artificial Incubation 33

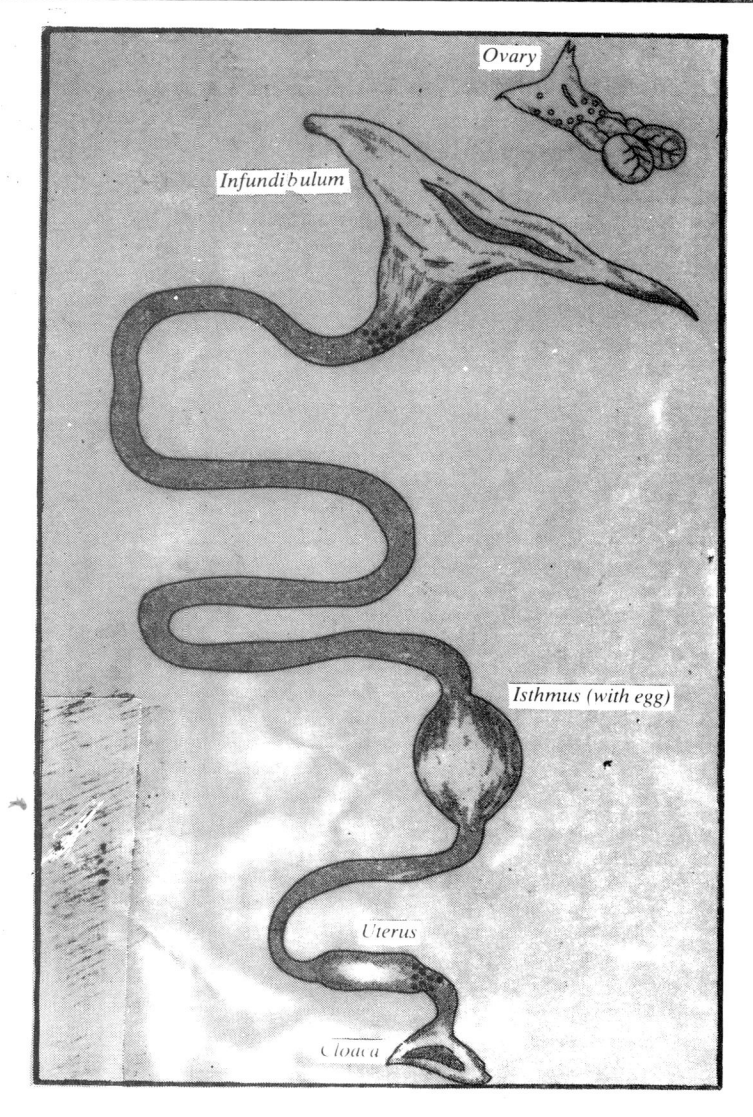

Figure 3.2 The Ovary and Oviduct (see 3.1)

and the spermatozoa which has moved up the oviduct into the infundibulum. From the time the egg begins its passage down the long tube to laying takes about 24 hours. This is why a hen cannot generally lay more than one egg per day. Each stage takes place in the following sections:
1. **Infundibulum.**
2. **Magnum.**
3. **Isthmus.**
4. **Uterus.**

The layers of the albumen are put around the yolk and so are the **chalazae fibres** that hold the yolk in position. At the end of the period the egg is turned and then laid. The approximate times taken are shown in Fig. 3.1. (Figures are taken from *The Avian Egg*, Romanoff and Romanoff.)

The oviduct is small when not in use and in the domestic fowl about 6 ins., but in full production it is around 2 ft. or more and much wider; the 'volume' may be increased as much as 50 times.

If the egg has been fertilized the creation process now starts. If a male bird has not been running with the hens, or he is unfit or immature so he is incapable of producing the necessary sperm, then no chicks are possible. Sometimes a cockerel is bullied by the hens and this has a negative effect on reproduction.

4
THE REPRODUCTION PROCESS

Breeding

The male bird fertilizes the eggs by 'treading' the female. In more technical language there is an act of copulation between cock and hen when the former ejects semen into the vagina of the latter. The spermatozoa goes into the oviduct and at the top one 'unit' of sperm links up with the ova and the germ spot is fertilized.

When the female is laying or about to lay she may crouch, with legs bent and wing shoulders raised and spread out to receive the male. If not, when the male makes the appropriate move to mount the female, and grasps the neck just below the back of the head, she will crouch and take in the semen. After the process is completed, taking only seconds, the female shakes herself and moves away from the cock.

In approprate circumstances **Artificial Insemination** may be practised, but this is *not* recommended for the fancier and even for a commercial venture it is a process which requires skill and dedication to detail. For birds which do not breed very readily, such as Indian

Game and turkeys, the semen can be collected into a glass tube and then the hen is held so that the oviduct is *everted* and the liquid transferred. Birds can be tamed to become accustomed to the process, but it is very labour intensive.

Development of the Germ (Blastoderm)

From this point **there is development of the germinal disc** which is started about 3 hours after the fertilization. An understanding of what is involved should lead to better results from the use of incubators. **The main purpose is to go along with the dictates of nature and not against the normal forces which have developed over thousands of years.**

This disc is on the yolk of the egg. Before fertilization it is called a **Blastodisc**; once fertilized it becomes a **Blastoderm** and is the most important part of the incubation process. If an egg has a section of shell removed it will be seen that the blastoderm, larger than a blastodisc, rests on top of the yolk and it is from this point that the embryo develops over the incubation period.

The process is rapid and, within two days, the embryo is well on its way towards being a living organism. In fact, at twelve days it breathes and creates heat.

Sometimes the two periods are distinguished by reference to the 'semi-living' period and the time when a chick is a live embryo as follows:
1. Aquatic period.
2. Normal Life period.

This helps to understand why the temperature, ventilation and humidity requirements have to be adjusted between the two periods and why the mixing of eggs in different stages of incubation is not advisable, although it can be permitted provided care is taken. **Thus it is not advisable to introduce new eggs when chicks are hatching because opening the incubator at that stage can be quite dangerous to the chicks attempting to emerge from the eggs.**

Stages in Development

The various stages in development should be understood so that each can be recognized when the eggs are candled or broken open when a hatch has failed. There is a definite pattern which can discerned throughout the period and any variation will call for action. The stages are shown in Figures 4.2 to 4.4 in this chapter. These consist of drawings and photographs to show what has occurred when dealing with the domestic fowl. When other species are being considered it will be necessary to

38   Artificial Incubation

adjust for a longer or shorter incubation period.

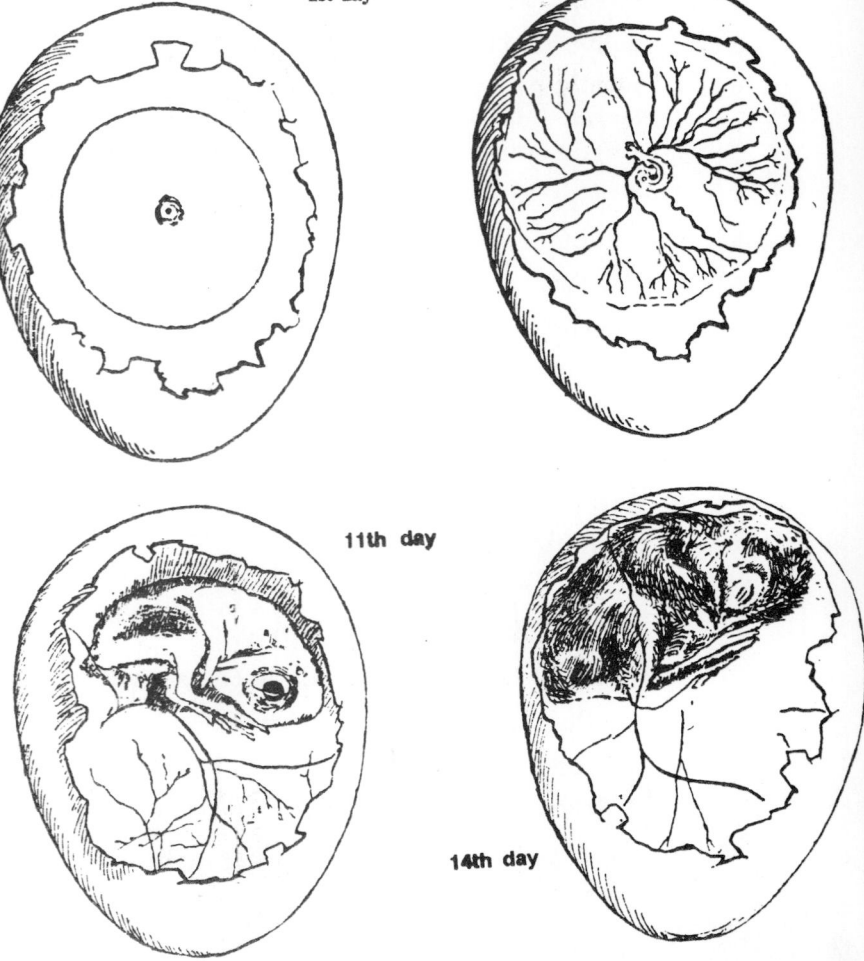

Figure 4.1   Stages in Development of Embryo

Artificial Incubation 39

4-Days

7-Days

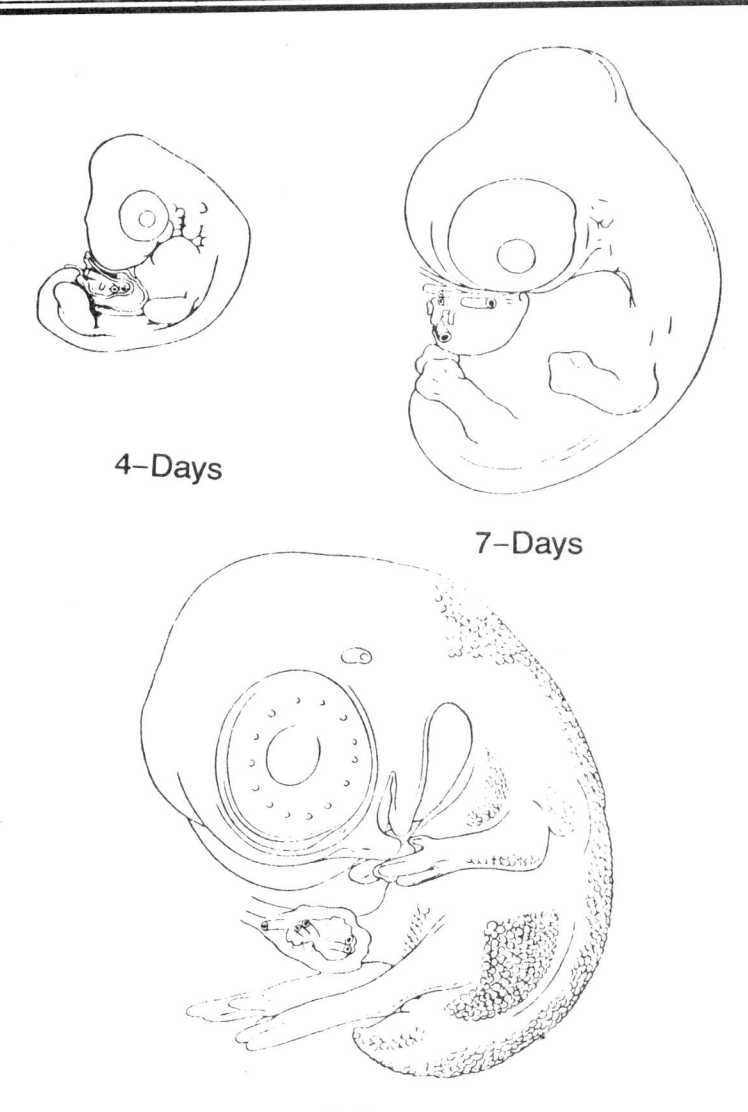

8-Days

Figure 4.2 Stages in Development (cont).

SUMMARY OF DEVELOPMENT OF CHICK

Day	Description of Change and Embryo Development
1 & 2	Across blastoderm the embryo develops and splits; chromosomes develop; centre grows, eyes, muscles and heart start.
3 & 4	Beating of heart, two blood vessels; wings, brain and head begin to be develop.
5 & 6	Blood circulates; liver, lungs and stomach now distinguishable.
7 & 8	Appearance of intestines, veins, beak and brain in ; flesh forms.
9 & 10	Ribs and gall bladder formed, beak completed; first movements.
11 & 12	Skull formed, orbits of sight, and ribs perfected.
13 & 14	Spleen and lungs now in position.
15 & 16 /17	Embryo now completing process into chick and the finishing touches are completed.
18 & 19	Chick now moving and starts to chip at shell to start to emerge. Yolk starts to be absorbed.
20 & 21	Hole appears in shell at broad end and chick extends until top of shell removable for chick to escape. Takes about 6 to 10 hours to emerge.

Notes: *The stages are approximate and there is overlap in functions.*
Not all developments are completed on day shown.
The process has been greatly simplified to aid understanding.
The description is the normal development of the domestic fowl.

Figure 4.3 Stages in Development of Chick

The Final Stages

The moment has arrived when the great escape has to be accomplished and it is now up to the chick to produce the energy necessary to get out of the egg. At this stage, if successful, it emerges from the shell all damp and weak from its exertions. The absorbed yolk should give it all that is necessary for the next 2 days, but thereafter there must be food and water. The procedures to be followed are described in the chapters on rearing.

If chicks cannot escape from the egg some breeders try to assist them. As a general rule this is *not* to be recommended because it means the chicks are too weak or the hatching process is not functioning properly. The ostrich chick is sometimes helped out of the shell because this is very tough and thick, but even there it should be possible for the hatch to take place without assistance.

Unfortunately, when there has been mismanagement or the eggs are simply not hatchable (a percentage come into this category) the whole process has been a tragic waste. A post mortem to establish the reasons is essential, thus learning by the previous mistakes.

Until the chicks are moved to a permanent home for rearing they should be left to dry and not disturbed in any way. However, do not leave in an incubator more than 24 hours because the temperature will be too high. A **rearer** will provide the more natural temperature requirement (*see* Rearing section).

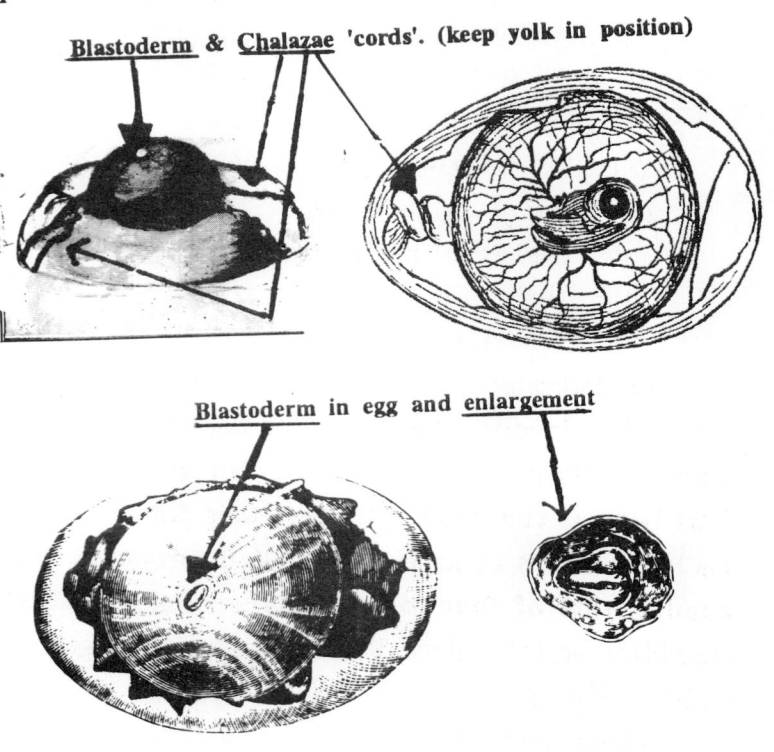

Figure 4.4 Blastoderm Illustrated
Top : Egg opened to reveal blastoderm and chalazae. *Bottom* : Egg with section of shell removed to show yolk and blastoderm.

5
HISTORY OF INCUBATORS

Objectives of Chapter
The purpose of this chapter is to show the reader how incubation has fascinated inventors since first recorded history. It also examines the various methods and means used to copy Nature, which is vital to an understanding of the principles involved.

Early Developments

As indicated in Chapter 1 the process known as 'incubation' involves the means of giving all the conditions necessary to develop the embryo so that it hatches. This book is concerned with providing the means artificially by means of an incubator. **It involves the correct combination of four key factors – heat, ventilation, humidity and turning.** These are shown diagrammatically in Figure 1.1 in Chapter 1.

Throughout known history attempts have been made and quite successfully to create the conditions needed to hatch eggs and this chapter traces the developments.

Early Incubators

The earliest attempts at artificial incubation occurred in Egypt and China. An early description of this process is described as follows:

> The art of hatching chickens has long been practised in Egypt. About 386 ovens are set up in various villages, each in the hands of an expert who is expected to hatch at least two thirds of the total and he keeps any in number which exceed that figure. There are about 92 million chicks hatched in these ovens.
>
> (*The Wonderful Magazine,* c. 1780, adapted)

This process required great skill and much training to hatch the **target** of more than the required two thirds of the total set, even in Egypt. In a different climate, one which is variable, with great extremes, the task would be extremely difficult, if not impossible. The effect of cold temperatures at night, hot spells during the day and variations from one season to another affect the relative humidity, ventilation and temperature. Accordingly, there is great difficulty in achieving a steady temperature over the full period of the incubation period.

Apparently the secrets of the Egyptian incubation processes were known only to the people of the village of Berme and a few adjoining places. The main essen-

Artificial Incubation 45

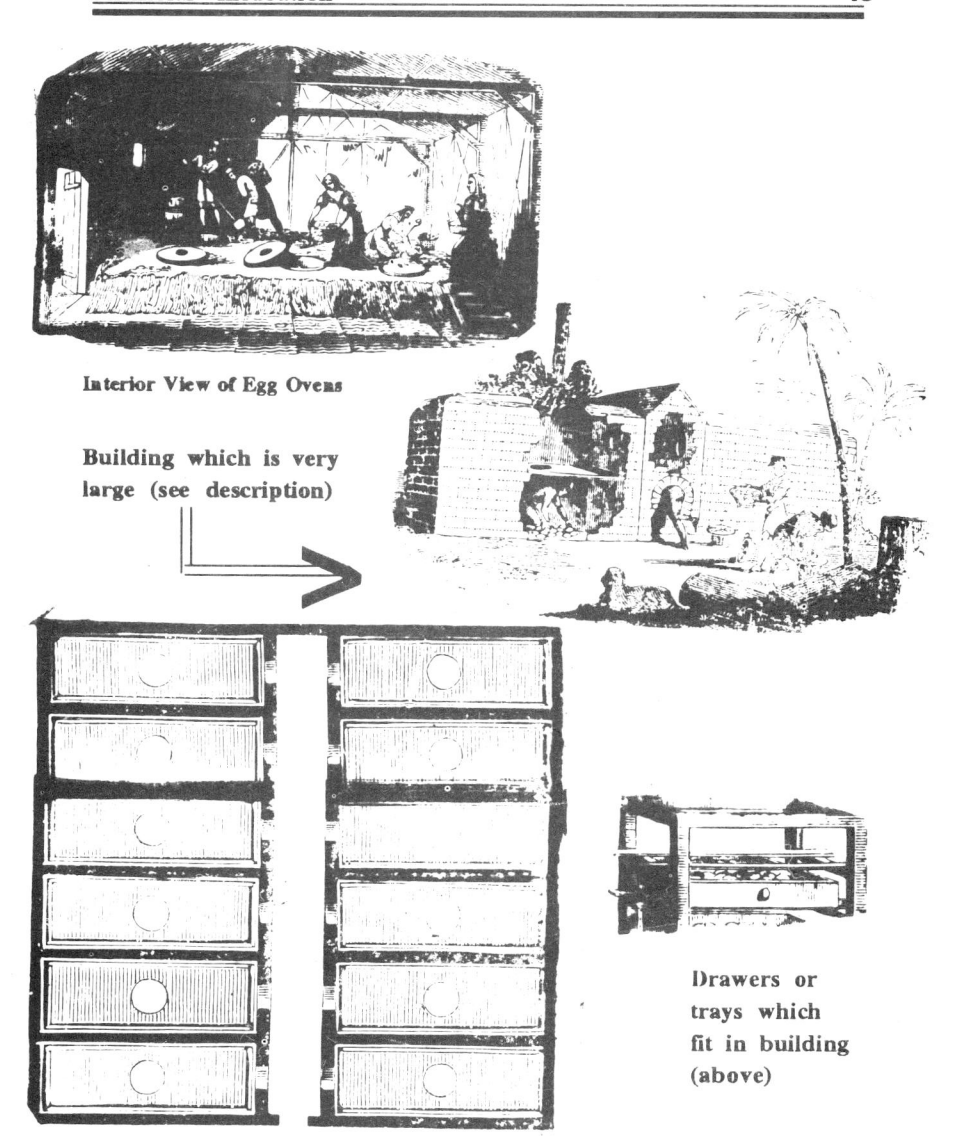

Interior View of Egg Ovens

Building which is very large (see description)

Drawers or trays which fit in building (above)

Figure 5.1 Egyptian Egg Ovens & Details
From old drawings (not to scale)

tials were as follows*:
1. Brick oven of 9ft high with a gallery in the middle 8ft high and 3ft wide which forms the entrance to the incubator.
2. Rooms at each side of the gallery in a double row, each 3ft high, 5ft wide and 12 or 15ft in length. Each has a round entrance for a man to go through. Each holds 4,000 to 5,000 eggs. The total number of rooms varies from three to twelve.
3. Eggs are laid on a mat or bed of flax or other conductive material.
4. Heat is provided by means of a fire placed in an upper room to heat the room below. The fuel used must be slow burning and therefore dried 'cakes' of dung of camels or cows, mixed with straw, are used. The fire heat is directed downwards and the oven is allowed to heat up so that at about 10 days (the exact period varies depending on the weather) the fire is let out. At this stage the oven is expected to have sufficient heat to complete the hatch. When the smoke has subsided the openings into the gallery are stuffed with bundles of coarse tow, thus confining the heat and keeping the temperature steady.
4. Eggs are moved into the upper rooms for hatching thus giving them better heat and more room.

All this was done under licence from the Aga of Berme (a sort of Franchise) and out of the total eggs (45,000) each incubator controller had to produce 30,000 eggs. In addition he received a sum of 30 or 40 crowns, his keep, and any surplus chicks.

---
* From various sources especially *Museum of Animated Nature*, Charles Knight, c. 1890

Artificial Incubation

The Hatching Room

The Brooder

Figure 5. 2 Reaumur's Inventions

Scientific Approach to Incubation (Reaumer)

The first attempts to hatch eggs in a form of incubator were carried out by M de Reaumur (*The Art of Hatching & Bringing up Domestic Fowl*, Paris and London, 1750). In this book, which is quite substantial, the author describes the Egyptian ovens and their operation.

He also gives the results of attempting to hatch by various means such as heating by manure which failed because of the fumes, then was successful after placing the eggs in casks sunk in the manure but raised above the surface about 3 inches. Next came the use of a baker's oven and the stoves used to heat greenhouses. Again, reasonable results were obtained.

He also recognized that the incubation temperature required was applicable for all types of birds, which meant that once this could be obtained the only variable was the differing periods to allow for the various species. For example, a hen would require 21 days and the turkey 28 days.

In his experiments, he allowed fresh air into his stoves to regulate the temperature and also used melted butter to show the heat being achieved – the state of the melting showed the approximate temperature level. With experience it was possible to achieve the correct temperature. On the turning and moving to different parts of the hatcher this was also done, thus imitating the hen.

The experiments also extended to artificial rearing and in this connection he invented brooders in the form of boxes which he called artificial parents. Some of the boxes he lined with fur or sheep skin to keep the chicks warm, gradually allowing them to be exposed to cooler air in a heated room until they were old enough to be let out. Later brooder were more sophisticated (See Fig. 5.4)

Early History

Whether the Egyptians were the first to have artificial incubation is difficult to state with certainty, but they showed people in other lands how the incubator ovens could be operated.

Some writers have suggested that the Chinese were certainly in the field very early in history. Records show that these were used on a large scale (10,000 eggs) around 200 BC.

Finding a Regulator

The main setback to the general adoption of artificial incubation was the means of controlling the temperature so that it could be kept at the desired level at all times. There had to be feedback within a system and automatic adjustment when required.

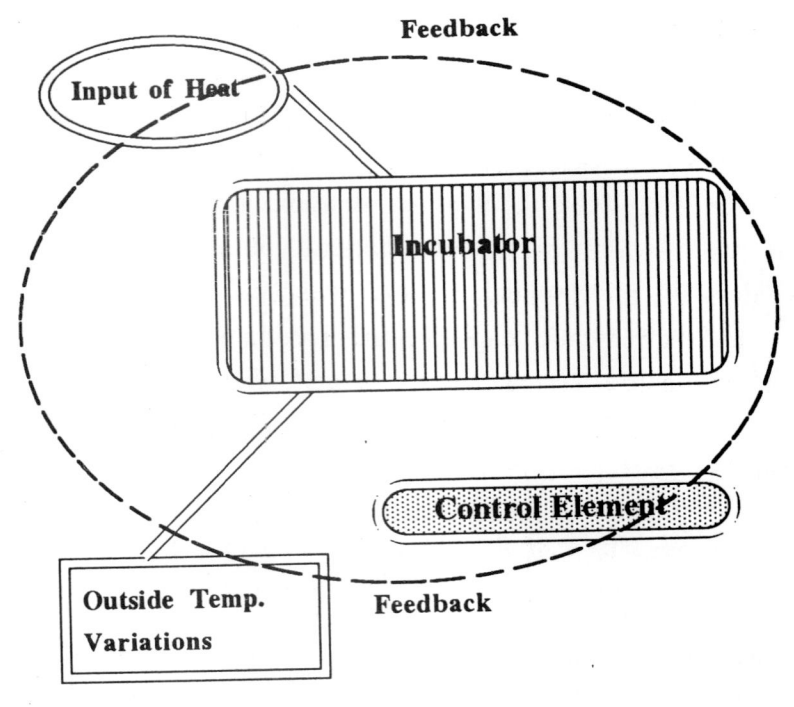

Figure 5.3 Feedback in the Incubator
The means of adjusting the temperature must be positive and reliable.

Charles Hearson & the Capsule (Thermostat)

The incubator has always attracted great attention and a variety of methods were tried to achieve the feedback required to regulate the temperature. There were clockwork devices, thermostatic bars, electric currents and even expanding air chambers, all to little avail because they were unreliable.

Charles Hearson recognized the problems and came up with a realistic solution in what he termed the **thermostatic capsule.** This was operated in a hot-water machine in which there was a tank which had to be heated up to the desired level by means of an oil lamp which ran at a fairly low level to avoid any smoking. Inside was a water tray which was half filled with water at 80 deg. F , and topped up twice weekly.

The thermostatic capsule, the feedback and controller within the system, was a round, flat capsule in which liquid was sealed. It was described as follows (from *The Problem Solved* by Chas. E Hearson) :

> ....the regulation of the "Champion" (incubator) is effected by putting about 20 drops of liquid (which boils just below the temperature which we wish to maintain in the egg drawer) between two pieces of thin metal which have been previously soldered together at their edges................

..as soon as the warmth is great enough to vaporize the liquid, the two sides are distended just as an india-rubber cushion does when blown up.

The invention, which occurred in 1881, made mass incubation a practical proposition and its effects were felt all over the world. The rise of the great enterprise in Ostrich Farming* and commercial poultry keeping on a large scale now became feasible. Yet the concept, like many other far-reaching inventions was quite simple. The capsule was fitted into the machine and a lever in the centre operated a mechanism which lowered or raised a 'damper' which was fitted on top of a chimney (See **Fig. 5.5 The Champion Incubator**).

The details of this machine have been explained because they serve to illustrate the general principles of the incubator. There are still incubators in operation which use this device. Obviously, though, an **electronic control** can be set to operate within closer limits and with less trouble, although when a capsule is used in an electrically heated machine these go on for years without trouble. I have operated one machine for about 25 years without any difficulty.

See *Ostrich Farming*, J Batty, **available from the publishers.,**

Figure 5.4 Front of *The Problem Solved* (see text)
This was dated 1905 and was in its 25th Edition!

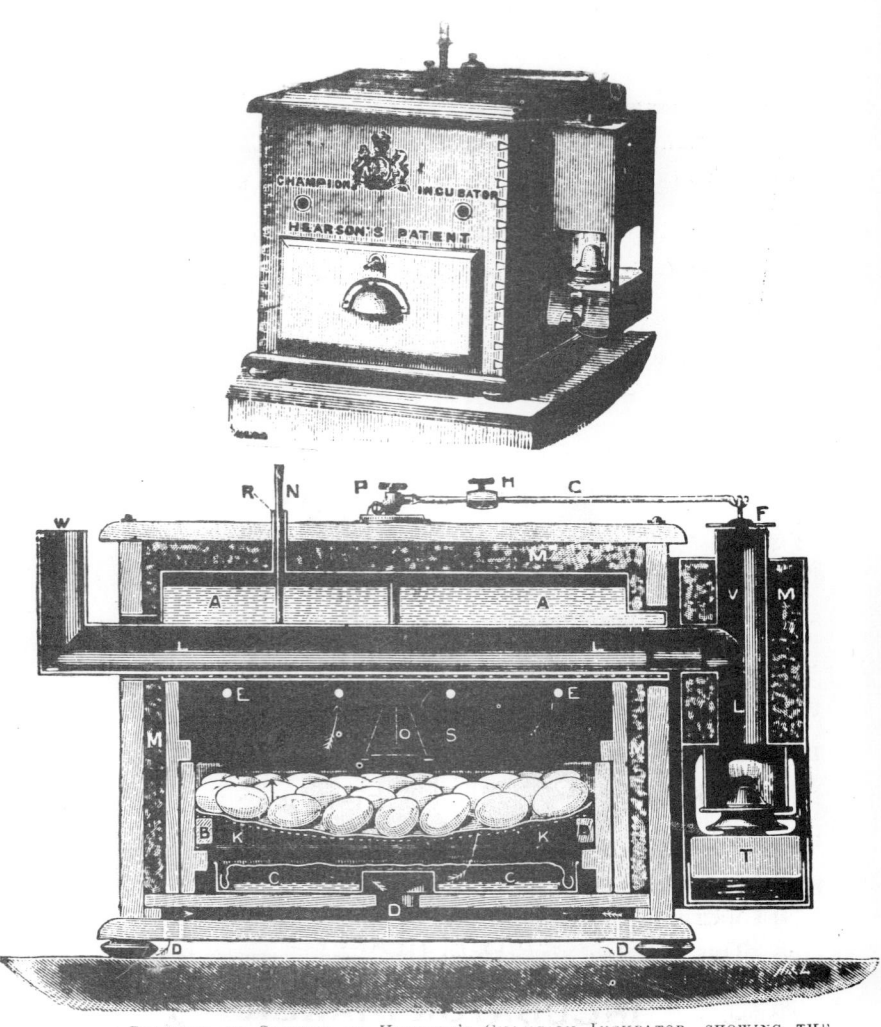

—DRAWING OF SECTION OF HEARSON'S CHAMPION INCUBATOR, SHOWING THE INTERNAL ARRANGEMENTS.

AA, Tank of Water; BB, Movable Egg-tray; CC, Water-tray; DDD, Holes for Fresh Air; EE, Holes for Ventilation; F, Damper; G, Lever; H, Lead Weight; KK, Slips of Wood; LLL, Lamp-chimney and Flue-pipe; MMM, Non-conducting Material; N, Tank-thermometer; O, Needle for Communicating the Expansion of Capsule (S) to the Lever (G); P, Milled Head-screw; R, Filling Tube; S, Thermostatic Capsule; T, Petroleum lamp; V, Chimney for Discharge of Surplus Heat; W, Chimney for Discharge of Residual Product of Combustion.

Figure 5.5 The *Champion* Incubator

Artificial Incubation

According to Hearsons, at the time, theirs was 'the only incubator in the world' which provided for different elevations above sea level; the normal capsule was for up to 1,000 feet above sea level. Altitude does affect the efficiency of an incubator.

Other British Incubators

Numerous incubators have been invented and many have failed as the quest for perfection continues. The Hearson incubator relied on a hot water tank which, once heated (which took a considerable time), acted as a safeguard against any failure of the heating system. Other machines relied on hot air and were quicker; more attention was paid to certain features and attempts were made to build in more safety factors. There are too many to list all the machines and therefore only the most notable are mentioned.

1. Improved Monarch Incubator (Wm Calway of Sharpness)

This had two drawers which used a water tank to heat and control the regulator by means of a float in the chamber which raised or lowered a case around the lamp wick, thus regulating the flame and the temperature. A further feature was the high level of insulation from packing with inch-thick hair felting, thus protecting the eggs for a considerable period if the lamp went out.

Artificial Incubation

The *Monarch* Incubator

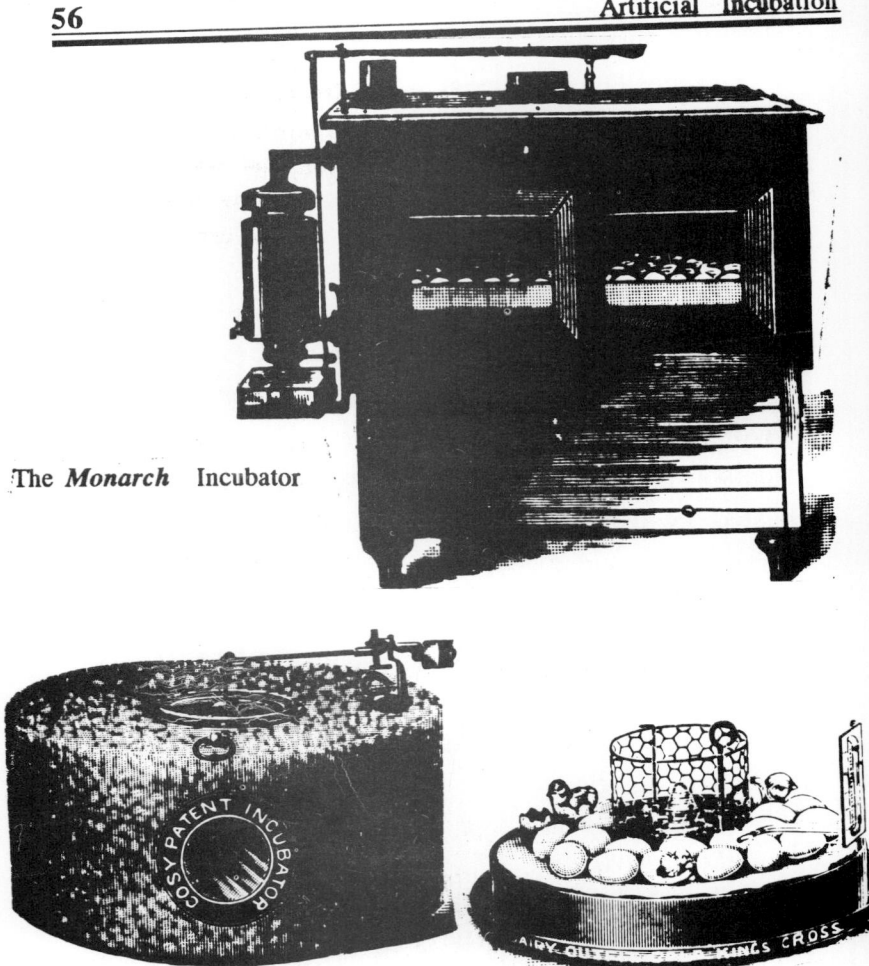

The *Cosy* Incubator (above), opened to turn eggs.

Figure 5.6 Various Machines (See Text)

Artificial Incubation 57

2. Nonpareil Incubator (W Taplin & Co)

Similar to the Hearson incubator, but the heat was supplied by radiation from hot water tanks which extend the water to the flue and lamp chamber, thus giving more direct contact with the flame. A capsule regulator and rod which were covered in insulating material, regulated the heat.

3. Cosy Incubator (Miss Wilson–Wilson)

This was the forerunner of many small, circular machines, being made for 30 eggs. It was also a hot air machine, but with a water tray at the base for humidity control. The heat came from a lamp which reflected the heat on to a metal plate at the top. A regulator was included and the thermometer could be seen from outside the machine.

THE NONPAREIL INCUBATOR.

THE USA EXPERIENCE

Claims were made in the USA that a self-regulating incubator had been found, also in 1881 the date of introduction of the Hearson Capsule; whether the two were connected is not clear. Many different incubators were made around this time. Charles A Cyphers built a mammoth machine, capable of holding 20,000 eggs; this was the start of large scale hatching.

The 'Cyphers' Hot-Air Machine

As noted, Cyphers invented a hot-air machine which became a practical reality in 1896. A regulator was used to control the temperature and it was claimed that the air was diffused in such a way that no extra moisture was required.

The thermostatic regulator was made of steel and aluminium and when the hatching temperature was exceeded the aluminium expanded making a connecting rod rise, which, in turn, would let some of the heat escape to bring the temperature back to the correct level.

Whether the device was entirely successful is not clear, but it would appear that the Hearson capsule was more acceptable because this continued to be used, and is still in use on older machines. It had the great advantage of being compact and could be fitted by the breeder without difficulty.

Artificial Incubation

Other Models

There were many other models invented, each with its own features which were said to be better than the competitors.

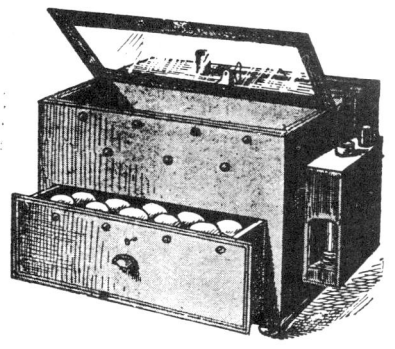

Taylor Incubator

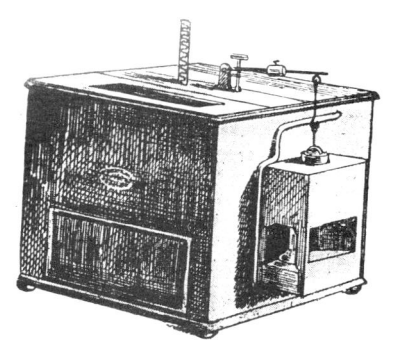

Nelson Incubator

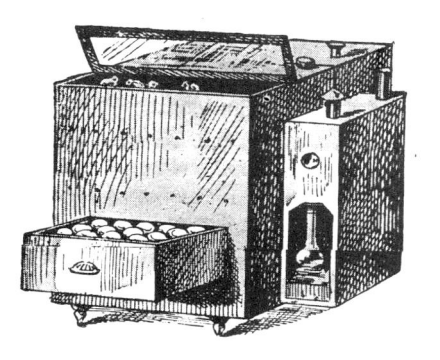

Thompson's Excelsior Incubator

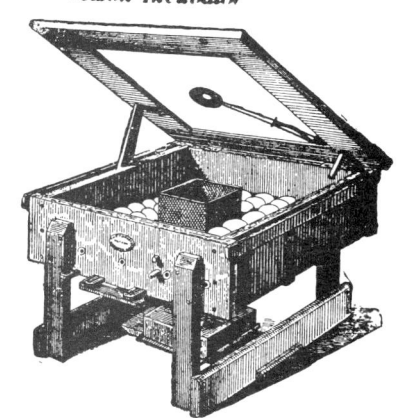

Forester Incubator

Figure 5.6A Various Machines (cont)
Names come and go in the quest for the 'perfect' incubator.

The Buckeye and Petersime Incubators

Early incubators for commercial use were introduced by **Ira M Petersime** (1923) who used electricity, and the **Buckeye company**. Both companies are still operating today and have developed many new concepts in incubation.

We have had no experience with the early Petersime, but do possess a very early BUCKEYE which introduced many new features for control, ventilation and humidity. It is operated by electricity which greatly facilitated the control. The many advanced features include:

1. **Mercury switch to control the temperature.**
2. **Fan to create draught through the incubator.**
3. **Special hatching section.**
4. **Device for turning the trays of eggs.**

All this is installed in a very large cabinet made of polished wood. Unfortunately, its sheer size means that it is no longer economic to operate. Later models were even larger and very complex, but improvements were effected for each new model.

Stoves had to be employed to provide the heat and then gas and, finally, electricity. The latter is controllable and therefore more suitable than any source which fluctuates.

Artificial Incubation 61

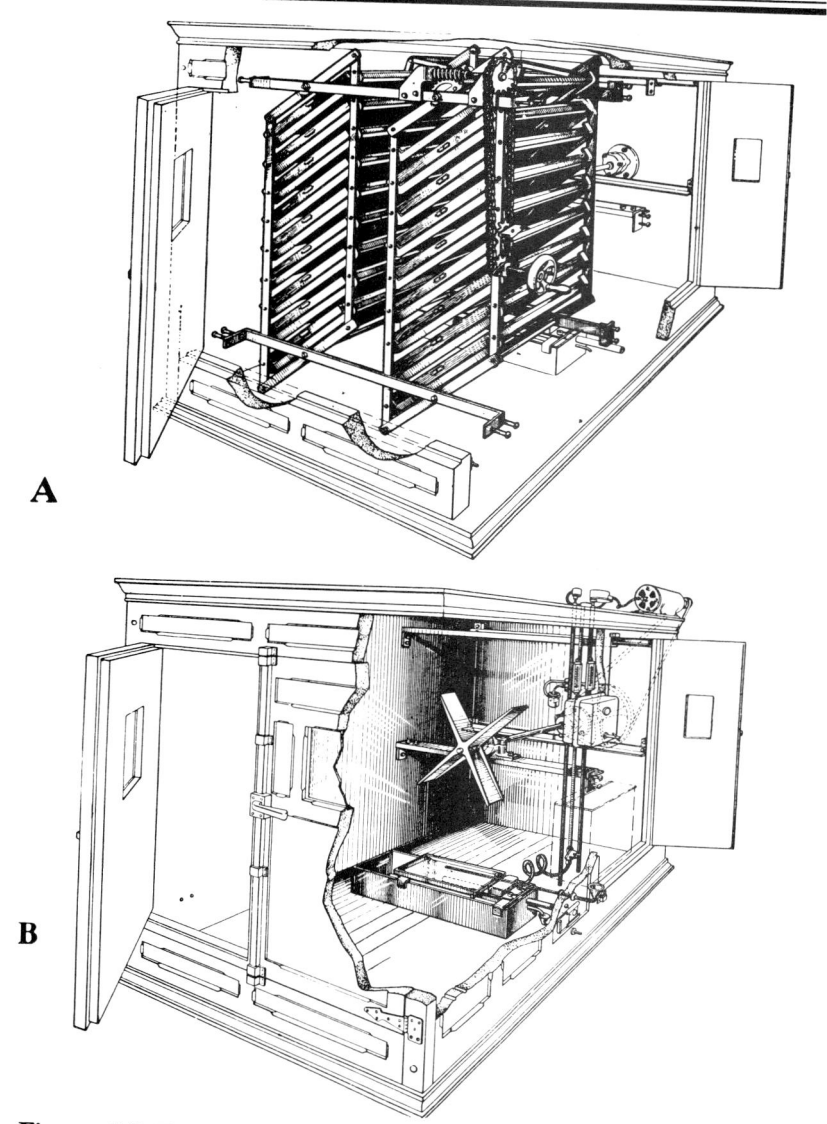

Figure 5.7 Examples of the Mechanism of Cabinet Incubators
The Turning mechanism (A) and Ventilation and Humidity Controls (B)

Smaller Modern Incubators

There were many new incubators introduced around the 1930s many of which operated by oil lamp. Usually they consisted of drawers which could be pulled out for filling with eggs. The oil lamp, later convertible to electriciity, was housed in the chimney with a capsule inside the machine to operate the damper mechanism. One of the major problems was to generate enough heat to maintain the temperature, especially in cold weather. A heated incubation room was certainly essential or the fluctuations would be too wide. Some users modified the insulation on the machine and this made them more economical.

In effect these were an improved version of the Hearson incubator and it was some time before the smaller versions of incubators came on the market which could be operated economically for a relatively small number of eggs. The *Vision-type* of incubator became very popular where the users such as schools could observe the eggs through a perspex cover and thereby provide a very practical lesson in biology. Similar machines are now produced which use electronic devices for controlling the temperature and other requirements.

In the USA, Al Marsh introduced various small incubators, some of which included automatic turning. (still being used).

Artificial Incubation

Automatic Regulation

The introduction of the capsule by Hearson was a landmark in incubation. There were many imitators and modifications. Their main purpose was to raise the damper on the chimney thus letting out the air and cooling the heating chamber. The capsule expansion was the main method, but there were others which included a metal strip which expanded and another, known as Rankin's patent which worked by expansion of the water in the tank (see Monarch Incubator earlier).

The essential features of a regulator are shown below, a version invented by J B Everall.

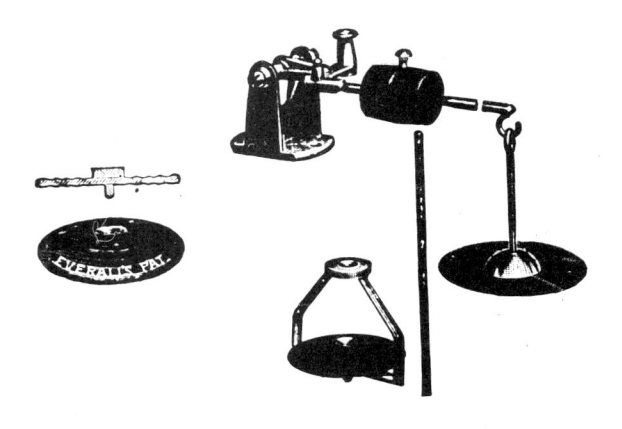

Figure 5.8   A Capsule Regulator

Candling

The checking process at various stages of incubation is of vital importance because it allows infertile eggs to be removed. **More importantly, the first candling shows whether fertile eggs are being produced.** There is nothing more frustrating than waiting 21 days for eggs to hatch only to find they were not hatchable anyway. Whatever is wrong, such as a cock which is infertile, can be spotted and corrected before much damage is done.

Candling is the process of putting the eggs in front of a bright light so that the contents can be seen sufficiently to decide whether there is an embryo; this is done at 7 days or other suitable period, as necessary, when the embryo shows distinct developments.

Various devices have been used for this purpose and some of these are shown below. Essentially a box and a light, shining through a hole is the main requirement. Some eggs are more difficult because of colour (brown eggs) or because of very thick shells (eg, Guinea Fowl eggs) and more powerful lights may be needed.

Which ever method is used for candling the aim will be to see whether the air space is shrinking at the appropriate rate and, if not, an attempt must be found to

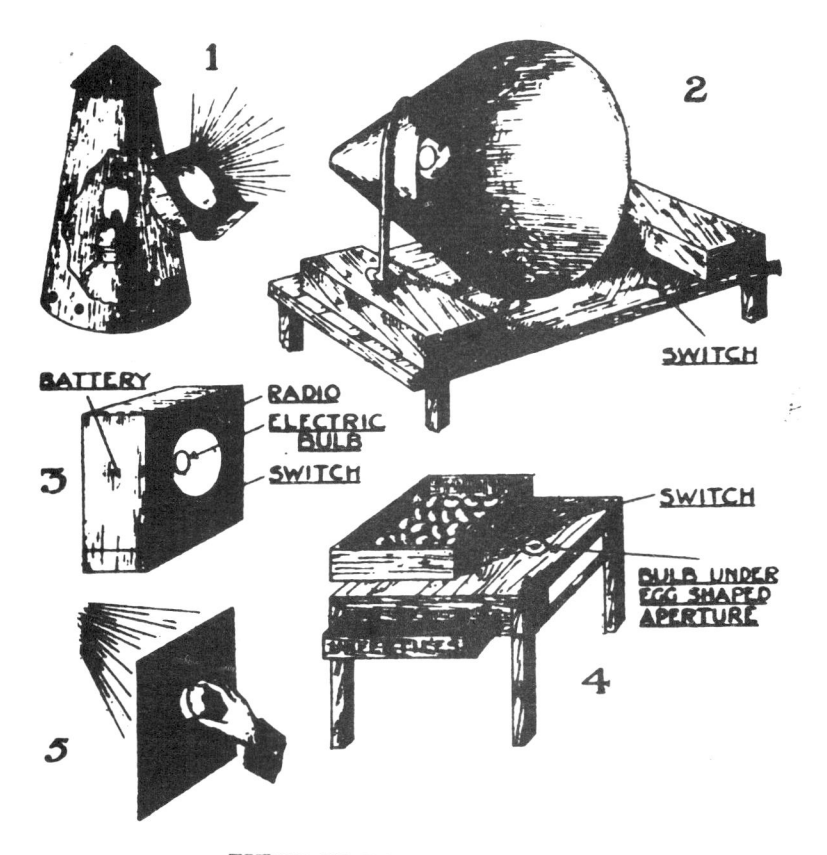

TYPES OF EGG TESTERS.
1. Usual type of Egg-tester Paraffin Lamp.
2. Large black cone, with small electric bulb at back, largely used for testing market eggs for freshness, spots, etc.
3. Small tester, with small bulb and dry battery. Radic points or marks indicate the place to put the egg.
4. Table tester, with space for egg drawer.
5. Sun through shutter of darkened room.

Figure 5.9 Various types of Egg Testers Used in the Past. A modern tester can be purchase or the above adapted and made by the handyman.

correct the fault. The 'clears' can be used as dog food or for some other purpose, but 'addled' eggs should be buried or destroyed.

From early times the need to investigate the possible reasons for failure was recognized and this process still continues today. Records should be maintained to ensure that improvements can be effected.

A Modern Development

The most significant modern development in artificial incubation has been the use of incubators for hatching the eggs of parrots and other rare birds. In the parrot field mention should be made of Rosemary Low and for exotic birds generally the directors of Bird World, Farnham, who have led the way in the techniques. Considerable work has also been done in Canada and the USA. Usually a small forced-draught machine is used because this avoids wide variations in temperature and maintains a more acceptable level of humidity than with the still-air machine.

Conclusion

Incubation has a long and interesting history. We can learn from the mistakes of others and gain from their example. It is a process which requires care and dedication as shown by the efforts of early pioneers.

6

JUSTIFYING THE EXPENDITURE

Essentials of Incubator Choice

Incubators are available in a wide range of sizes and with a great variety of features which help to make the machine more reliable or easier to operate. Generally the size is the determining factor in making the decision; **this is also linked with the capital expenditure involved (price plus installation) and the running costs of operating the incubator.** In the case of a very large machine a 3-phase electricity supply may be very desirable.

In some ways the large machine, once mastered is easier to operate than the small machine because the former has many more automatic controls and safeguards. Moreover, the large machine, capable of hatching thousands of eggs in a single 'setting', requires a professional, management approach because any mistakes may be extremely serious.

**The basic approach is the same irrespective of the machine and no book (except a machine manual for the specific incubator) can supply the necessary technical knowledge to operate an individual incubator, because each one is different.

The Golden Rule, therefore, is to follow the instructions supplied by the manufacturer.

Committing yourself to a particular incubator means that after considerable study and a comparison of the relative merits of each you have selected that machine to cover the purposes for which you require it.

A comparison may involve the compiling of a Check List to see whether one machine is **better** or **worse** than another. An example is shown opposite and this may be modified in the light of specific requirements. It will be seen that the selection should be either X with 65 points or Z with 68 points. The price will probably be the final deciding factor, but the possibility of technological change and the known reliability of the manufacturer must also be considered.

Do not jump to the conclusion that a machine is better because it is advertised more or the makers make exaggerated claims.

Many machines have been used for many decades and they should be examined very carefully before going for a brand new concept which has not been tried and tested. As shown earlier, the history of incubators is filled with machines that are no longer available.

Artificial Incubation

	CHECK LIST		
	Short List of Possible Incubators		

Requirement: Still Air Incubator to hatch 200 eggs each month.

Selection: After study of market made a shortlist of 3 machines, X, Y and Z.

	X	Y	Z
COST	£500	£600	£1,200
Features: (Mark out of a scale of 10)			
Self Turning	6	6	8
Humidity Control	5	8	9
Ventilation	7	6	7
Visible Hatching	8	8	8
Electronic Controls	9	3	9
Materials Used in Construction	Wood 7	Plastic 7	Fibre/G 9
Expected Life Span	5 yrs	5 yrs	8 yrs
Country of Origin	UK 9	USA 7	Italy 5
Parts/ Maintenance	9	9	5
Comparisons:	65	59	68

Note: The relative costs and 'points' are shown to illustrate the principles involved and not to make a comparison between actual machines.

Justifying the Purchase

When dealing with the small poultry person, whether this is a fancier or somebody who makes a partial living from hatching and selling special stock the choice and justification will probably be based on the hope that it will pay for itself in a few years. **Overall cost will probably be the determining factor and whether funds are available at the time.**

No commercial poultry farmer can afford to install a very large incubator costing, say, £100,000 in the hope that it will eventually pay for itself. If it is a matter of obtaining replacement pullets each year then outside purchase may be the most economic way; large scale incubation is an operation which requires great skill and management so as a start in the justification the basis must be that there is a sound commercial reason; eg, supplying pedigree stock, show birds to fanciers, pullets to poultry farmers or broiler chicks to farmers. There may even be a very specialized venture such as hatching ostrich eggs or guinea fowl eggs for a known market. **The Business Objective must be recognized at the very start or the venture is unlikely to succeed.**

In terms of commercial justification this may be accomplished by basic facts such as working out the expected sales and measuring these against the total

Artificial Incubation 71

```
Date. 17/8/19                            No. 362
                   Capital Expenditure Request
From: FARM MANAGER

To:                       Secretary, Capital Budgeting

Details of Proposed Purchase:
   XL4 INCUBATOR, 10,000 EGG
```

Costs Involved:	£
Price to be paid	x x x
Carriage	x x
	x x x
Installation:	
Material	x x
Labour	x x
	x x x

Reasons: TO SUPPLY DAY OLD BROILER CHICKS TO CONTRACT

Expected Benefits: ADDITIONAL REVENUE £50,000 PER ANNUM (COST SCHEDULE ATT'D)

Date Required: WITHIN 6 MONTHS.

Year of Payment: 19

Approved/Rejected

.... Divisional Manager

.... Managing Director

Figure 6.1 Capital Expenditure Proposal

costs over the life of the incubator. A variation is to calculate how long it will take for the machine to pay for itself. Generally there should be an expectation of a **payback period** of 3 to 5 years because beyond that time, the incubator could become obsolete and would most certainly have a greatly reduced value in the books; remember that each year an incubator would depreciate and this would be charged in the Profit & Loss Account.

The methods used to decide on whether investment in fixed assets (those kept in the business for use with a life in excess of one year) are known collectively as Capital Expenditure Decisions and are used to compare alternative assets and to justify whether purchase is worthwhile. This is a very large subject on which many books have been written and therefore readers requiring further details are advised to study a specialized work.

The methods used are as follows:

1. Payback Method

If an incubator is to cost £10,000 and revenues from selling chicks over 4 years are £1,000, £2,000, £3,000 and £4,000 respectively with £5,000 expected in the fifth year then the Payback period is 4 years. Alternative propositions would be compared and the one with the shortest payback would be selected.

For example: *Management Accountancy*, **Batty J,** where forms and procedures are described.

2. Return on Investment

This approach examines different proposals and calculates the percentage return expected on the investment. In the above example under the Payback method earnings of £15,000 were expected over 5 years and this is a 50% earning rate.

An alternative approach is to state the earnings for each £1 invested so that comparisons can be made quite easily. Thus two machines may have an earning power of £1.50 and £1 respectively when it is quite clear on the financial measure alone the first incubator would be selected.

In using any of these methods it is advisable to use the net amount for the earnings flow; otherwise a false picture would be portrayed for one machine may incur much higher running costs.

This approach may be modified by taking the return on the investment, known as the **Return on Capital Employed method.** A percentage return would be specified and this would be used as a minimum rate of return for all proposed investments.

This may be a figure of 20% although the precise figure taken would have to be determined for each business and would be influenced by the state of trade generally; in a recession a lower figure would have to be accepted.

3. Present Value Rate of Return

Instead of taking the figures at face value the expected returns are converted to *present values*. This means that an amount expected in, say, 3 years is worth less than a sum expected within one year. It is the concept of earning money from an investment put into reverse. Thus the cash flow of £1 expected in 3 years is worth today only £0.7513 at a rate of 10 per cent; in 5 years each £1 would be worth £0.6209.

The problem with the conventional methods above (1 to 3) is that they ignore the **timing of the receipts** and therefore do not show the true values.

Basis of Comparison

Listing the possible incubators and comparing their relative merits is made easier by the systematic approach suggested above. Moreover, the discipline imposed ensures that all possibilities are explored before a decision is made.

Some agricultural economists in the past have suggested the best way of calculating the machine to select is to work out **the cost of producing a specified number of chicks;** *eg, say, £2 per 100 chicks*. This figure is purely hypothetical because comparative cost figures only become meaningful when taken at a specific period of time and for specified conditions. However, the general principle is sound and quite simple to understand, whereas the **Present Value approach** may be difficult to understand by the practical poultry farmer.

Artificial Incubation

Figure 6.2 Large Incubator where trays are stacked or wheeled in, on special trolleys.

THE INCUBATOR BUILDING (HATCHERY)

The selection of a room, or in the case of large scale hatching, a building will be of vital importance. If an existing building is available this might be adapted when the additional costs would have to be considered. If the building is being used for another purpose then to arrive at a realistic assessment of alternative projects it will be advisable to calculate the loss of revenue from its change of use or any possible alternative purpose. This is known as an "opportunity cost" – the calculation of the lost benefit from taking the building for another purpose.

In some circumstances a decision may be taken to go ahead with modifying a building even though in the long run a special, purpose built building would be more efficient and cheaper to run. The determining factor might be the capital available at the time; if insufficient funds are available the cheaper approach might be necessary.

In these circumstances, where capital is the limiting factor, the possibility of renting a building or getting someone to finance the cost and taking it on a leasehold basis might be the answer. However, if at all possible, it would be preferable to own the premises because this gives control and provides security for possible future borrowing.

THE INCUBATOR ROOM OR BUILDING

This chapter examines the important question of suitable accommodation to house the incubator(s) and other essential equpment. The correct choice or adaptation can make all the difference to the success of hatching chicks. Irrespective of the size of the operation the same principles apply. As noted in the preceding chapter any expenditure should be justified before work on the building is started. There is thus a working plan, linked with the Business Objectives.

CONSIDERATIONS FOR A SUITABLE BUILDING

An unsuitable room or, in the case of a hatchery, a badly designed building or collection of buildings, can make all the difference between success or failure. The fancier who tries to hatch eggs in an incubator located in, say, a bedroom, or the poultry farmer who uses an old poultry shed out in a field, may both be fighting the elements more than the problems which arise from the incubator itself.

The fancier should place the incubator in a sound building without windows and, if possible, a ventilation cowl, which can be opened at different levels or closed. Insulation is absolutely vital and should be done properly so the temperature remains steady even in the summer. Generally hatching will be done from December on, but usually in the Spring when room heating will not require to be very great. The most effective is to line the walls with insulating board or special polystyrene insulating sheets (taped at the joints), which can then be decorated.

A brick-built building is best, but expensive. Wood or other material may be used, but not corrugated sheets which act as heat conductors.

Inside, attention should be paid to the walls, which should be decorated with a non-absorbent material. If washing down is to be carried out some form of laminated board may be ideal. Avoid any material which attracts flies, insects and vermin.

Because new materials are constantly becoming available it is advisable to seek expert advice before proceeding and remember that a top-class building and insulation will pay for itself very quickly in terms of improved results.

Artificial Incubation

Detailed Planning

A **solid foundation** is essential and if a brick building, or pre-fabricated one, the floor should be made of concrete to a depth of at least 1 foot thus eliminating the scourge of the poultry man – rats – which seem to find away into a building even when this seemed impossible. If along the lines of a modern farm building made with a framework structure of steel to which are bolted sheets, then the girders should be set well into the ground. Where there is a great deal of traffic from road or rail, ensure that vibration is not excessive to the extent where it will affect the eggs.

Make sure that the **exits are adequate** for the traffic to be used and there is no free access for anybody to walk into the building or germs may be carried. The **question of security** and vandalism must also be watched for stealing from farms seems to be one of the curses of the modern age.

Preferably there should be double doors and **a safety porch** to avoid the air getting into the incubator room straight from outside. The provision of facilities for cleaning and fumigating incubators is also essential.

If the operation is planned very carefully it should be possible to **separate the incubation and hatching stages** so that there is one room for each, which can be kept at the appropriate temperature, thus saving on heating. This is because the hatching room, dealing only with live embryos, can have a lower heat level and higher humidity.

The separation also allows for more effective controls to be exercised over the **cleaning and fumigating of eggs** and ensure that new eggs are not contaminated with any germs that might exist in the eggs which are hatching. Some overlap will be inevitable, but the mixing together of different batches should be kept to a minimum.

In effect the large hatchery is like a laboratory which should be controlled in every way to avoid spreading disease; the conditions created are intended to maximize the development of embryos ; unfortunately, this is also conducive to the spread of any germs which might be latent in the eggs so rules on hygiene must be formulated and strictly followed.

In designing the building attention should be paid to the **flow of the eggs through from the 'intake', washing where necessary, fumigating, traying-up, transferring to the incubator, moving to the hatchers, taking out the chicks and putting them in boxes ready for despatch.**

In modern buildings provision is made for **isolation**

Artificial Incubation

tion areas and tunnels through which eggs are transferred to give the necessary washing or fumigation. Even if not carried to its ultimate of fumigating every egg some form of sterilization will be advisable. Dipping in a mild disinfectant, specially formulated for eggs, will reduce the risk of infection and this is now practised by many hatcheries.

The **fancier** can disinfect the eggs quite easily by using warm water in which the disinfectant has been added.

If *formaldehyde* is to be used this must be prepared and administered under controlled conditions because the gas is dangerous. This comes in different forms and the strict instructions of the makers must be observed.

Formalin, which is formaldehyde in a liquid form, is quite popular. Another chemical, **potassium permanganate,** is added, in a ratio of 2 parts to 3 parts of formalin. A mixture of 8 oz. should fumigate 100 cubic feet of space.

There are sprays and a special type of chemical formaldehyde which can be activated by heating on an electric plate heater.

The formalin treatment to eggs at the beginning of the hatch takes half an hour. Thereafter the gas should not be allowed to come into contact with the live eggs or the chicks when hatched.

The provision of protective clothing, including gloves, goggles and mask is an essential reqirement.

Floors

As noted earlier, floors are best made of concrete and covered with a waterproof (non-slippery) "skin" which is washable. This allows regular washing down and ensures maximum hygiene. If a brick building is used the walls should be plastered, preferably with a water resistant material so they can also be washed (*see* below on insulation)

In the interests of absolute cleanliness there is much to recommend the use of wall tiles, especially where washing or other operations are carried out.

Artificial Incubation

SIZE & LOCATION

The size of the room or building should be determined by reference to the **number of incubators and other machinery and equipment required to carry out the function**. It will be necessary to calculate the floor space required and the height of the building by reference to the tasks to be performed. Generally a height of *less* than 8 ft. (say, 3 metres) is not to be recommended, but excessive height should be avoided because the extra space still has to be heated.

Corridors and partitions should be planned to give maximum utilization. A layout drawing is advisable to show the optimum arrangements. Starting with the intake the trolleys will have a planned route through the system.

If an underground incubator room is to be built the depth should be about 1.5 metres below ground and the remander (1.5 m) above ground level. If on a split level basis (built into a hill side) the access side can be above ground.

Factors Influencing Size

The equipment needed to produce a throughput of chicks, keets, ducklings or other stock would have to be calculated by reference to the list given overleaf.

ESSENTIAL EQUIPMENT & ACCOMMODATION

1. **Incubators** to produce the weekly output required throughout the year. Obviously the normal maximum need should be provided so that flexibility is possible.

As appropriate, separate **setters** and **hatchers** should be provided in adjacent rooms, but separated to maximize heat utilization. Leave enough corridor/room space for access and dealing with the normal running of the incubators.

2. **Trolleys and Special Trays** for traying and loading and unloading the incubators.

The equipment is designed to minimize the handling costs, allowing all tasks to be carried out on the plastic trays which are washed and sterilized after each usage.

If the laying sheds are located a long distance from the layers (desirable to avoid spreading of disease) **a van or other vehicle may be necessary to move the eggs and the chicks.**

3. **Egg Washers to cope with the throughput.**

Not all agree with washing eggs and prefer sandpapering or other method of cleaning, but the fact remains that this process may be essential, especially when hens are running out.

4. **Tables or benches** for storing and sorting eggs and traying them.

5. **Candling Lamps** or related equipment for checking the eggs at prescribed intervals.

6. **Generator** as a standby in case of a power cut.

Artificial Incubation

7. Table/ equipment for sexing chicks when this is to be done on the premises.

8. Trolleys for stacking the chick boxes as chicks are taken from the hatchers.

9. Room for storing the eggs and getting them ready for putting in the setter once per week.

10. Any other special rooms, cupboards, shelves, and washing equipment needed for maintaining a high level of hygiene in the hatchery.

All this will be converted into square and cubic metres or feet and then made the basis of the planning.

REQUIREMENTS

The requirements for suitable accommodation to house incubators are as follows:

1. An Even Temperature

2. Adequate Ventilation

3. Humidity at the Correct Level

4. Access and provision of any necessary services.

Even Temperature

The heat in the room will have a direct bearing on the operation of the incubator at the correct level. If the temperature falls very rapidly in the room then there must be a compensatory rise in the heat provided in the incubator. If the variations are quite substantial there is a danger that the machine will not be able to cope and the hatch will be affected.

The normal room temperature should be about 60 deg. F.(15.5 deg. C.). A few degrees either way will not matter, but wide fluctuations are to be avoided. A central heating system may be used fuelled by gas or oil with radiators in suitable places – preferably not very near to an incubator.

When the incubators are operating the heat from them will provide some of the total heat requirement and this should be allowed for when calculating the number and size of radiators. However, if winter hatching is to be practised the radiators and boiler must be adequate to maintain the ideal temperature in all weather conditions.

If humid conditions are to be supplied this will have the effect of cooling down the temperature. At one time it was fashionable to suggest having a cellar basement for the hatchery building and this has many advantages. The ground level can be used as an office or warehouse whereas the basement can house the incubators, as well as providing a washing room and storage faxcility. Care has to be taken to ensure ease of access for eggs and chicks will have to be moved on a regular basis.

The interior should be thermostatically controlled so that heat fluctuaions are avoided. Insulation, using the special materials available for poultry houses which give minimum losses from the building, should be used; eg , "Styrofoam" from Sheffield Insulation. In this way the environment is quite controllable.

Adequate Ventilation

Fresh air should be drawn from outside at a steady rate. Some installations have fans whereas others rely on ventilators. Air is taken from a point near the floor and drawn through the room by the use of a cowl sited on the roof. A shaft and opening into the room draws the air as necessary; some form of regulator can allow control to be exercised over the rate of flow. Insulation of the roof space, and, indeed, the whole building, will be essential or the air will not be controllable in an effective manner.

In the very large building it will be necessary to install fans or even air-conditioning which may be the most economic in the long term. However, the cost will be quite high, both in terms of capital expenditure and running costs (although it may be less expensive than a piecemeal system which is not as effective). Accordingly, each possibility should be examined before a decision is made. Much of the secret of achieving the correct temperature and humidity is to provide a very effective level of insulation, and incorporate the ventilation into the design of the system.

Humidity

Humidity in the hatching room will affect the level in the machine so attention to both will be essential. In some cases no extra moisture may be necessary, but, in others, there may be some factor which dries out the atmosphere with the result that a humidifer will be needed. We are all familiar with the container of water fitted to radiators in a domestic situation where the air is improved and excessive drying out of furniture is avoided. It is a similar principle in the incubator room.

A great deal also depends on the type of incubator. Those which have a moisture tray fitted can supply more than enough for the room and the hatching process. There should be a steady level of humidity and this should be conducive to the machine producing an acceptable level of hatching.

The humidity is certainly affected by the temperature – one affecting the other – and if ventilation is too rapid the air may dry up and reduce the humidity, which, in turn, will increase the temperature. This will call for an increase in humidity, but since this will tend to reduce the temperature an adjustment will be essential.

If the cowl or baffles used for extracting the air are open too wide there will be too rapid a circulation of air with consequent changes to humidity and temperature levels. Getting the correct balance is of vital importance.

Artificial Incubation

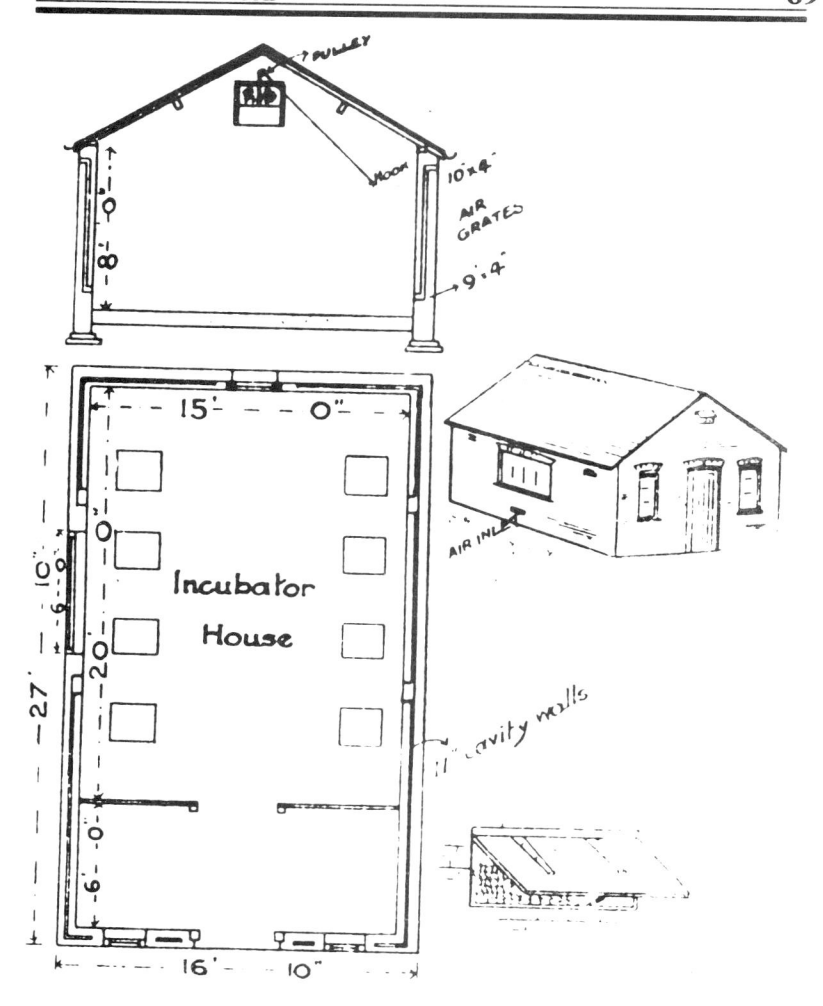

Scale 8 feet to 1 inch

Figure 7.1 Incubation Building
Intended to show ventilation openings. The 'windows' are covered
Adapted from a plan drawn by Will Hooley, FZS; FBSA.

Provision of Services

An hatchery is rather like a factory. Raw materials come in at one end and the finished 'goods' go out at the other. In planning the layout think along these lines and an efficient unit should result.

Various principles have been formulated from many generations of practice and experience. These are as follows:

1. A high level of hygiene is essential.

It is necessary to provide for regular washing and sterilization of trays and equipment between each use. Accordingly, have sufficient equipment to cover this requirement and have the facilities needed.

2. Adequate lights should be installed with special lights at table level for dealing with eggs and chicks.

The concentration should be on the working areas not over the incubators.

3. Storage should be available for items purchased in bulk such as chick despatch boxes which are overprinted with the name of the hatchery.

A year's supply of boxes would ensure maximum savings on unit cost.

Artificial Incubation 91

Figure 7.2 Trays for Storing Eggs.
Drawers which are given labels to show pen from which eggs taken.
The idea can be used by the fancier or the smallholder with limited stock with incubators which do not use trolleys with combined trays. (see Fig. 7.3)

4. Egg storage facilities should be as ideal as possible.

Opinions vary on the exact temperature and humidity for storing eggs, but the temperature should be neither too high nor too low. A steady level of about 55 to 60 deg. F. and relative humidity of 70 is the usual requirement, but a cooler temperature is permitted. If too high the eggs will deteriorate more rapidly.

The eggs, collected on a regular basis (four times a day preferred) are washed and/or 'sanitized' ready for putting in the setter. It is usual to use a marker pen to show the breed or pen and the date of laying. Some breeders advocate marking with a 'X' on one side and a 'O' on the other side (or similar marking scheme) which facilitates the turning of eggs. This is certainly useful for the **fancier**, but is not likely to be practical for the hatchery set-up where many thousands of eggs are being dealt with.

Special trays may be employed for stacking the eggs and in large incubators **these are the basis of the flow system because they are wheeled into the setter and subsequently to the hatcher.**

Opinions differ on how best the eggs should be stored. Following nature they should be laid flat in sawdust or similar material. However, provided they are not jolted too much, the yolk is kept in position by the challazae; in any event the yolk will always 'rise' so that the

Artificial Incubation 93

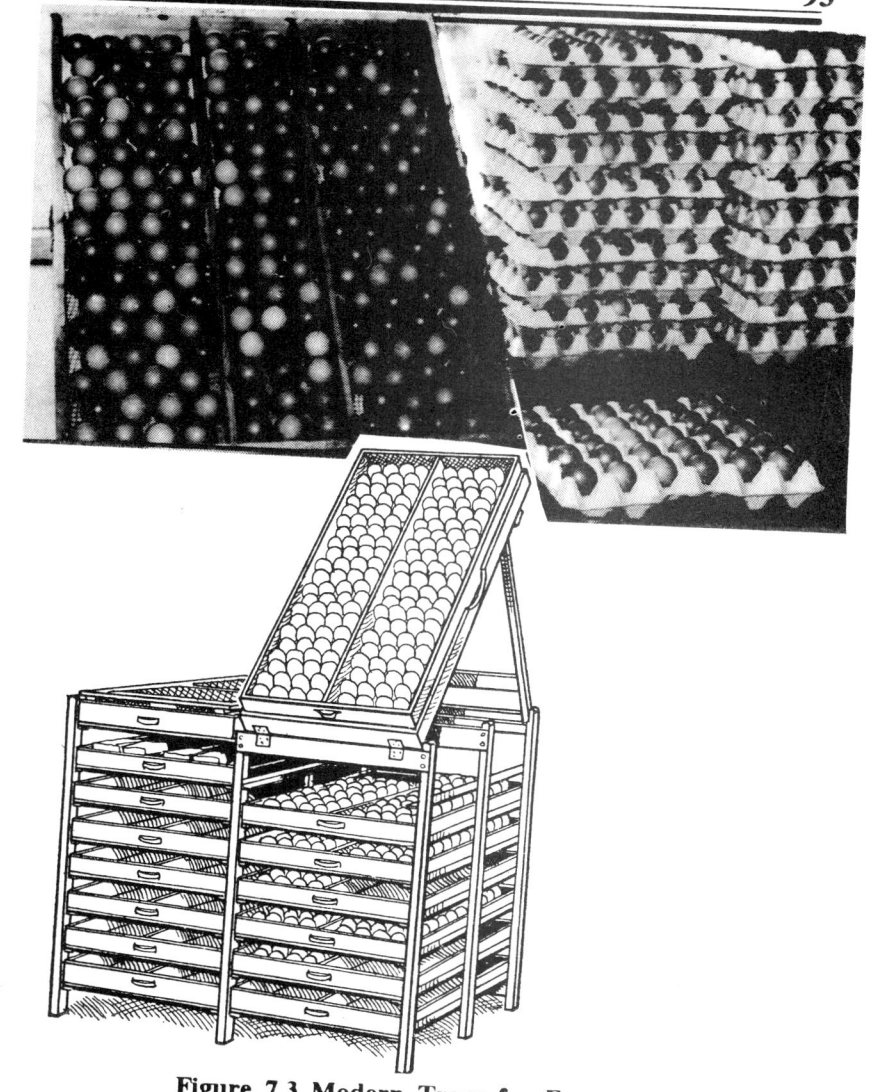

Figure 7.3 Modern Trays for Eggs.
These may be used separately or may be combined with a trolley which wheels into the incubator.

blastoderm is in the correct position. The fact remains, all eggs should be treated with great care to ensure they remain free from damage and not too hot or incubation will start prematurely. If a yolk bursts there in no hope of hatching the egg.

Keeping eggs free from draughts and from too much evaporation are also of great importance. In this connection it is advisable to use drawers as suggested or cover the eggs over with a clean cloth. **If the storage period exceeds four days there may be advantage in turning the eggs twice per day.**

The fact that eggs must be a typical ovoid shape and free from blemishes or rough shell has already been noted in the chapter that deals with the egg. After the eggs have been collected they should be marked and stored; this should be done daily. Then there should be a regular schedule for setting in economic quantities. The fancier may have difficulty early in the year because of shortage of eggs; also an incubator made for 200 eggs is not going to be very economic if run with 20 eggs.

If special, 'pedigree' eggs are to be hatched these should be placed in a section of the incubator reserved for this purpose. Sometimes small bags or a separate wire container are used to separate the eggs, but this is not likely to be very economical except for quite valuable birds when full details are to be recorded.

8
SMALL-SCALE INCUBATION

Basic Principles the Same
The basic principles will operate irrespective of the scale of operation. Therefore in this chapter the differences that do exist between small- and large-scale levels of operation will be emphasized. Each is trying to maximize the hatch of healthy chicks, those likely to live and grow into strong, healthy adult birds. Small machines were traditionally always the still-air type, but now forced-air incubators are also available. for a small number of eggs.

Getting Ready

The start of the season will depend on the type of birds kept and the objectives which the breeder is trying to achieve. Thus, for example:

1. Exhibition Purposes
Those requiring birds for the Autumn shows will aim to have eggs in December for hatching that month or January. Other fanciers may prefer to hatch in March or April because they can rear when the sun is shining each day and the outdoor temperatures are reasonably warm.

2. Heavy Breeds
Breeds such as Wyandottes, Orpingtons and Plymouth Rocks should be hatched early on so they are growing from March onwards to obtain size and start laying in the Autumn and through the Winter.

3. Light Breeds and Bantams

These should be hatched from the middle of March and in April. Again this should ensure maximum egg production during the Winter months. The large breeds are the layers such as Leghorns, Minorcas, Andalusians, Campines and Anconas. All bantams are best hatched in the spring-time.

These are examples to show that thought must be put into the best approach to adopt. Plan when the incubator is to be used and then make sure this is done properly.

Checking & Testing

Before the incubator is put on there are a number of matters to be checked on:

1. Check the room temperature to make sure it is at a reasonable level.

The minimum room temperature is 70 deg.C,(21.2 C), but may be between 55 and 65 deg. F. If too low there will be difficulty in maintaining a steady temperature in the incubator so a poor hatch will result. A steady temperature in the room means that the **self-regulating feedback** will not have to be implemented as much.

Some experts give a table to show the temperatures that should be present for a still air machine at given room temperatures (F); thus:

Room.	40	50	60	70	80
Incubator	105	104	103	102	101.5

Artificial Incubation 97

2. Run tests on thermometers and capsules if used.
An independent check should be made on any electronic regulator to make sure this is running properly. Capsules and thermometers can be checked in water of the correct temperature. If the machine is coming into operation in a new season make sure that proper conditions are operating before eggs are put in the incubator. *Once all is ready, run the machine as long as necessary (eg, 2/3 weeks) to make sure that the operation is running steadily for a period. Have a 'dummy' run at first.* Only when both temperatures -- room and incubator -- are running steady, will it be wise to start the incubation proper.

3. Wash and disinfect the machine and give the incubation room a Spring-clean.
Amazingly the last hatch of the season invariably results in a machine being left in a neglected and dirty condition. Egg shells, stains and even water dried in the moisture tray are all to be found when the new season starts.

This is quite wrong and, if possible, the machine should be washed down at the end of the season and then again when the new season starts because this keeps an incubator in better condition and gives it a longer life. It also sterilizes and kills bacteria. A small incubator may last 25 years or longer if maintained properly.

The cleaning involves the removal of all the sections and where they are galvanized metal, plastic or similar the sections should be soaked in a bath containing hot water, washing-up detergent and disinfectant, such as Jeyes Fluid. After about one hour they can be scrubbed and then hosed down with cold water. Once these have dried they can be put back in the incubator.

It is not wise to soak any wooden parts such as the frame or outer walls, but these can be wiped over with a cloth soaked in a washing solution as described above, but newly made.

4. **Collect eggs and mark them with the date of laying and an 'O' on one side and 'X' on the other to facilitate the turning process.**
At this stage a decision will have to be made on whether to wash and sanitize any eggs by dipping in a warm solution of water and disinfectant. Opinions differ on this matter, but it does seem to be beneficial to sanitize the eggs even though this does not generally give protection against germs already inside the egg.

OPERATING THE SMALL MACHINE

At this stage everything should have been tested so the eggs can be placed in the machine and the hatching started. If eggs have come from an outside breeder it is generally wise to wait a full day befor placing them in the machine; they should most certainly be disinfected.

A routine should be established so that tasks are carried out according to a set timetable. In this way mistakes are kept to a minimum.

As many eggs as possible should be placed in the machine following the manufacturer's instructions. For the newcomer to incubation this is the period of learning and hope. Despite what is stated by some suppliers, there is no such thing as a perfect incubator!

Artificial Incubation

There are many variables which affect how the incubation process will proceed and this is why so much depends upon making sure that the **controllable factors,** such as room temperature, are regulated so there are no wild fluctuations. The number of eggs placed in the machine will influence its behaviour and notes on the experiences at **different levels of filled capacity**; eg. *quarter full* , *half full* , will help to get better results in the future. Because of the better results obtainable an incubator is at maximum efficiency when almost full and this should be the aim.

Humans, acting rationally, and trying to be as economic as possible will tend to purchase a machine which caters for the maximum number of eggs at the height of the season. There are a number of reasons for this approach:

> **1. Prices of machines makes the larger machines more economic in terms of cost per egg hatched;**
> **2. The larger machine gives greater flexibility and tends to take up proportionately less space per 100 eggs than would a number of small machines;**
> **3. More features, including automatic turning, exist on the larger models.**

The result of this 'rational' decision may mean that the machine is running at under-capacity for most of the time.

This also results in the **cost per chick** being much higher than budgeted and the control is more difficult. A method of overcoming this problem is to half fill the machine at the start and then fill the remaining space after 7 days. Some may fill one-third each week and this results in a steady flow of chicks. However, if the incubator is both setter and hatcher this three-time-filling approach is unsatisfactory because of the conflict with the hatching period and, as noted earlier, the need to have a high temperature and humidity at the end of the incubation period. In reality, because of the heat from the chicks the incubator heat is lower and must be at this level for optimum results, but **in total** a high temperature must be maintained although not so high as to cause accelerated hatches with attendant problems. Thus the following situation arises:

Stage I Day 1, 100 eggs placed in incubator. Hatch, Day 21 (21st)
Stage II Day 7, 100 eggs placed in incubator. Hatch, Day 21 (28th)
Stage III Day 14, 100 eggs placed in incubator. Hatch, Day 21 (4th next month)

In these circumstances the ideal conditions will never be present because, from the twelfth day of Stage I, there will be live embryos and, when the twenty-first day is reached, the Stage III eggs will still be in the 'aquatic' stage which requires different conditions from those required for hatching.

A better approach might be better to avoid the Stage III filling and limit the loading to two lots of eggs each month. This is an alternative, but it must be said that some breeders claim that they can get acceptable results placing eggs in once per week.

My own experience suggests that the incubator should not be disturbed from the penultimate date of the hatch; ie, the 19th day for poultry, and it would be foolhardy to place a fresh batch of eggs in the machine at this critical stage. Even removing the newly hatched chicks should be a carefully controlled operation, being taken out in batches of a reasonable number, say, 10, and the top closed quickly afterwards so the air and humidity cannot escape. Otherwise, the danger of death after pipping is very real.

There are possible ways of overcoming the problem:

1. Wait until all the chicks have hatched on the 21st day before placing new eggs in the tray.

2. Place eggs in the tray which have already been *incubated by broody hens* **for a period of 7 days. Even here the timing should avoid the** *hatching period***.**

When there is a separate hatcher the problem is removed or at least reduced in size because it is the final hatching stage which can result in poor hatches.

Loading the Incubator

The way a machine is loaded depends on its type, but where a tray is involved it may be found to be more advantageous to work from the middle. If hand turning is necessary the eggs should not be packed too tightly or they may be difficult to grasp and turn. In addition, it is essential to ensure that the air, humidity and temperature reach the eggs. Avoid filling the corners of the square-shaped machine because the heat may not always reach the outer corners. In any event, when turning, it will be essential to move the eggs to different parts of the machine, thus imitating the hen when she incubates her eggs.

When eggs are removed after candling at 7 days it is permissible to add new eggs, but remember the principles relating to the conditions required by eggs of differing ages.

If an incubator has a fully-automatic turning device or one which is semi-automatic the loading will be influenced by the design. Where, for example, there are metal rings with oval gaps for the eggs, each one must be placed in a space. When the ring is turned the eggs

Artificial Incubation

turn and the larger machines have a number of rings, each smaller than just over the width of an egg, so all eggs can be turned simply and quickly.

Turning & Cooling

The eggs should be turned at least twice per day and for many small machines this involves using the hands (which should be quite clean). Large eggs such as ostriches or geese will require both hands and careful handling. Usually the hatch is quite satisfactory with two turns so if more are not possible this number will suffice.

With hand turning too many turns may result in the eggs getting cooled too much. If the *X and O system* is used the fact that a complete turn has been made can be seen quite clearly, although even a partial turn is quite adequate so long as the yolk turns and goes to a different part of the inner shell.

The practice of incubation described in older books always gave instructions on the benefits of cooling the eggs, thus copying the hen. In more recent times, and particularly with the very large incubators, it seems that special cooling is not vital. However, there is no doubt that the natural cooling given by the hen when she leaves the nest for feeding, does result in very strong chicks, although, clearly, there are other factors to consider, such as the ideal temperature, which is always present without much variation. It would appear therefore that in the small incubator the cooling given when eggs are being turned or candled should be beneficial. The fact remains that eggs must never be allowed to cool too much or the embryo might die.

Care and discretion must be employed in deciding when cooling can be permitted. In cold weather it may be better omitted altogether.

Watching the Temperature

The person attempting to hatch eggs must watch the temperature, humidity and ventilation on a regular basis. Moreover, there must be confidence in the machine and a full understanding of how it can be adjusted before committing a large number of eggs.

The following matters should be watched very carefully:

1. When eggs are to be loaded into the machine make sure the pre-loading trial has been successful. There should be no wild fluctuations or the hatch will not succeed.

2. Eggs can be placed in a cold incubator or one which is warmed up. Immediately they are placed in the trays the time should be noted so that a watch can be kept on how quickly it warms up to the desired temperature as discussed earlier. DO NOT TRY TO HEAT THE EGGS TOO QUICKLY.

As noted earlier, a trial period of running the machine is vital. Once it is running satisfactorily, *and only then*, are the eggs placed in the trays. *Try to anticipate the rise in temperature which will occur when the eggs are placed in the machine* . If run at between 101 and 102 deg. F then expect the machine to settle at 103 deg. F in about 2/3 days. If this is the recommended temperature with the thermometer just above the eggs (almost touching), then this is satisfactory.

It will take some time to warm up the eggs, and the exact time depends on the machine, the temperature of the room and various other factors. Be patient with this stage because, although it is not critical for the machine to reach the hatching temperature within, say, 3 hours, the process should not drag out too long. Neither should there be a soaring up of the temperature to a high level within half an hour because this is probably a sign the machine is running too high.

Do not expect adjustments to occur immediately the control knob is turned; some machines need at least 2 hours to respond. Some experts suggest that adjustments should be made at intervals of 8 hours. Be patient, but watchful and above all do not allow too large a margin of error to occur when the eggs are in the machine.

When the incubation room is cold (below recommended minimum temperature, especially on a night) then the possibility of further insulation must be considered. This could be a simple expedient such as reducing the airflow into the machine or placing a blanket or rug over the machine, but do not block any air holes, inwards or outwards, or the embryo may die. The timetable of the incubation will affect the air required, more being required when the embryos are alive and especially near to the day of hatching. However, on no account try to incubate when the room is so cold that it **feels cold** (eg, below 40^0 F.) because then the outcome is very doubtful.

Reaching a very high temperature for a few hours can be very dangerous and therefore must be avoided. If it does occur, cool down the eggs and hope the damage is not fatal.

3. Remember, once the machine is full, the controls will need adjusting to make sure these apply to a *full machine*.

Generally there is a simple regulator which can be used to turn the heat up or down. Do it in easy stages, but watch very carefully. Do not turn on the incubator and then go away for the weekend hoping all will be well. If you cannot afford to spend a considerable time checking the vital factors (see Chapter 1) in these early crucial stages then you should not try to hatch eggs this way.

4. If water is required in the machine then pour a reasonable amount into the water tray as instructed by the Instruction Booklet, but make sure the air regulator is set as directed (usually a low setting early on).

Some machines require very little moisture, but others need a constant supply; this is affected by weather and related conditions. It is better to go sparingly at first, watching the air space to make sure it is at the correct level.

The optimum position of the water in the water tray should be established and then watched very carefully. In some machines a fall in the level below the half way mark may cause a rise in temperature and

Artificial Incubation

\multicolumn{5}{c	}{**INCUBATOR CONTROL CHART**}

Incubator Type/ No.:...................... Breed/Pen...............

Date Set:............................ No of Eggs............

Date	Temperature		Humidity	Notes
	Room	Tray	Machine	

Results
Unfertile Eggs........
Broken Yolks........
Dead in Shell........
Pipped but Died......
Cripples................
Lost............%........

CONTROL INFORMATION (Targets to Achieve)
Date due:..........................   Eggs removed: No............Date:...................
Temp. Target:......................   Humidity..............
Room:.......... Tray:..................   Total Hatched:.................. %..................

Figure 8.1   Incubator Control Record
This may be adapted to suit particular needs.

it will be even more dangerous if the tray goes dry altogether. It all depends on the prevailing conditions. Sometimes eggs will hatch with very little water in the machine because the weather is warm and humid; at other times there must be no drying up of the moisture or the hatch will be spoilt.

The most dangerous situation I have experienced is when the machine has been run relatively dry and then at the 17th or 18th day the water tray is filled. This appears to cause the eggs to over react with the result that they absorb too much moisture and the chicks are drowned and fail to hatch.

The lesson to be learned is to run the machine under conditions which are as normal as possible and watch the air spaces very carefully, even weighing the eggs at key times so that irregular drying of the moisture does not occur. If there is too little moisture the air space will be too large and therefore the humidity should be increased gradually.

Where extra water is advisable such as for ducks, geese and fancy waterfowl the eggs may have to be sprayed with warm water.

Those eggs requiring less water than normal in the early days such as certain bantams, and pheasants, should be watched carefully so that the ideal air space is achieved. A drawing showing the **ideal** for each species should assist in deciding the level to aim for. Remember the total weight loss is about

Artificial Incubation 109

12% at the end of the period with a gradual decline to that level.

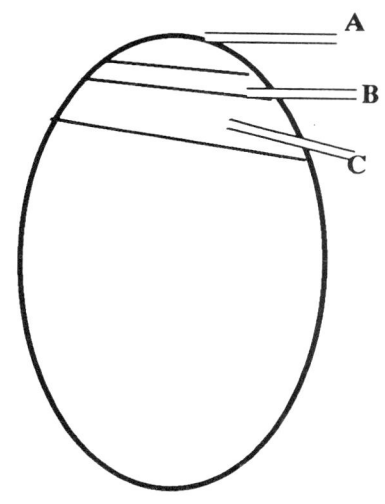

Note: The spaces are not drawn to scale or intended to show exact accuracy; each species will have variations.

Key:
A = 3% at 8 days
B = 5% at 10 days
C = 10% at 17 days

Figure 8.2 Weight Loss Indicator

Total loss at 21 days is 12%, but the reduction will take place gradually; eg, 3% at 8 days, 5% at 10 days, and 10% at 17 days. The loss depends on the species. The results may be plotted on a graph.

SELECTION OF MACHINE

As noted, the smaller machines , known as 'Table Models', tended to be **still-air machines** which usually required hand-turning, but many modern machines now have devices for turning eggs by operating some form of lever arrangement or by the machine being fitted with some form of turning device. Moreover, small forced-air incubators have been made available whereby the ventilation is achieved by means of a fan.

A sound, good quality machine should be purchased and although automatic turning is an advantage, especially for someone who is away all day, this should not be over rated; opening up the incubator and turning the eggs does give an opportunity to check the interior to make sure that all is well. Moreover, for those who believe in daily cooling of the eggs this is an excellent time to carry out the process. In fact, this period and that taken for candling will suffice for cooling.

The incubator should be of sound construction and strong enough to be lifted when full without fear of strain. Make sure the sections can be lifted out easily for loading the eggs, filling with water as necessary, and for cleaning. No materials likely to constitute a danger to chicks or operators should be used. The early incubators used to be made with asbestos sheets for strength

The Marsh Incubator

A Brinsea Incubator

Figure 8.3 Small Still-air Incubators

and insulation. This would no longer be permitted being regarded as a dangerous material.

The instrumentation should be capable of being seen and operated. An irritation is a thermometer that cannot be read unless it is taken out of the machine and examined in a particular light. Even with electronic controls it is wise to have spare thermometers and hygrometers to check that all is well. In fact, it may be wise to run thermometers at different points in the incubator to test that the eggs are being warmed properly; there may be one touching an egg and another 5 mm above another egg so that after allowing for the difference due to position (the higher one should show a higher temperature) it will be a safeguard against faulty thermometers.

Visual Type or Solid Top

The preference for the small incubator is to have a "see through" lid which lifts off when attention is needed and this normally has a padded cover which provides the insulation. The cover must be flexible yet contain the heat within the machine; even so it does tend to be a poor insulator compared with the solid top which might have a number of layers of insulating material. The advantage of the top being clear is that the chicks can be seen hatching. However, it should be apparent that if the

Artificial Incubation 113

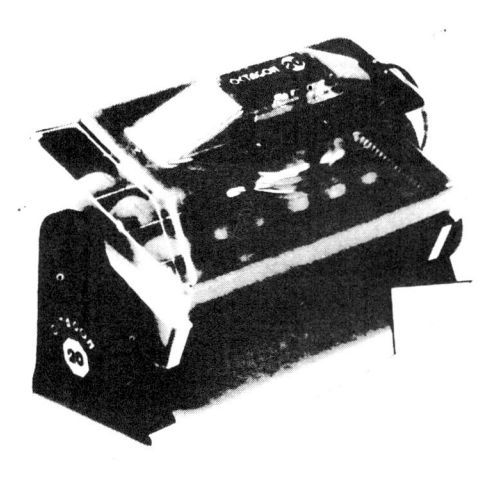

A Brinsea Incubator with Fan-ventilation and automatic turning

Vision Incubator (Reliable)

Figure 8.4 Typical Small Incubators (Cont.)

insulating cover is to be removed for observation, then the room must be quite warm to compensate for the loss of heat. Moreover, it is not a wise practice to leave off the cover because the machine will need to be adjusted to maintain the heat at the required level.

Some incubators have a limited opening for viewing and this overcomes the problem of loss of heat through perspex which cannot be insulated. Obviously though the viewing tends to be more restricted.

If the incubator is used in a school class room or at a zoological garden make sure that there is no tapping on the glass or attempts made by observers to remove the chicks before they are dry. Incubating eggs is a serious business and an excellent way to teach zoology, but the process should not be misused.

Separate Hatcher

Generally the small breeder has managed with the single machine which combines setting and hatching, but it is possible to obtain a separate (additional machine) for the hatching. This allows new eggs to be introduced into the setter at weekly intervals and 3 days before hatching is to commence the eggs are transferred to the hatcher, thus allowing the correct conditions to be obtained. Obviously, if the purchase of the extra machine

Artificial Incubation

is to go ahead there has to be justification in terms of regular use and space available.

Do not attempt to hatch too many chicks with a single machine; overcrowding only leads to problems. If very small eggs are to be hatched such as quail or extremely large eggs such as goose or ostrich the machine will have to be a special incubator or modified to suit the eggs. In these circumstances it would be advisable to consult the supplier of the incubator in question to obtain technical advice.

The storage of eggs has been covered in the preceding chapter. Remember the eggs should be clean, set before they are 7 days old and, after 4/5 days storage, they are better turned or be stored in trays which are tilted each day.

Special trays or a nest of drawers may be used for storing the eggs, but for the very small breeder a section of an egg tray made of compressed cardboard (in which eggs are purchased from supermarket) may be used, because these can be stacked and are quite strong. Make sure they are very clean and if eggs break then throw away the tray and obtain a replacement. Labels can be used to mark on any essential facts such as breeding pen and latest day for placing in the incubator.

Plan each stage for the best results.

Modern Refinements

Modern technology has enabled new devices to be introduced even into the small incubators. Whether these can always be justified is a matter for debate and there are some breeders who would argue that invention makes incubation less of a challenge and not as interesting. The fact remains that electronic devices lead to better control and cut out some of the risks. **Digital thermometers are now available and so are humidity control units which make for greater control.**

Keep it Simple

Eggs of different sizes and from different species can be hatched in the same incubator, but where the hatching requirements are different – length of period, humidity, and ventilation – it is better to keep them in a separate machine. In any case where there are marked differences in sizes it will be difficult to judge which is the top of the egg so the guideline on temperature is lost. The spreading of diseases from different species must also be considered.

Candling Eggs

It is advisable to have a definite timetable for candling eggs. Do not try too early because the embryo is not always clear in the first few days. For fowl eggs 7 days and 14 days may be best, but for **one candling only** a 10 day check would suffice.

9

LARGE SCALE INCUBATION

What is Large Scale?

There is no fixed rule on what constitutes 'large scale' or for that matter 'small scale' The latter can be as low as 15 hen eggs and as high as 500 eggs, but when the latter level is reached the incubation is approaching professional level and would be regarded by some as being a large operation. The size of incubators provide a guide line because once this reaches around 1,000 hen eggs they introduce quite sophisticated features, although as shown in the preceding chapters, even the very small machines have forced draught ventilation and automatic turning. The fact remains that around one to two thousand egg-sizes upwards would usually be regarded as coming within this large scale category.

Size to Select

The economics of investing in incubators is covered in Chapter 7, along with consideration of the hatchery. In practical terms, it will be necessary to estimate how many chicks are to be hatched and for what purpose. If the intention is to specialize in hatching pullets then it will be necessary to hatch in total about three times more than the output required; for example, 300 pullets will be obtained from 1,000 eggs. This will allow for cockerels hatched, failures in incubator and mortality of chicks. This target, whatever it may be, will determine the size of the machine.

There are also technical considerations which affect the decision. The small, still-air incubator does not work very efficiently for large numbers of eggs and around 150/200 eggs is probably the maximum number of hen eggs that can be managed. Ventilation and related problems limit the level of efficiency. **The optimum level of carbon dioxide which should be present is 0.40%** and once there is a build up beyond this level chicks may die for shortage of oxygen. If too low this will also retard growth. Yet for the specialist producer this type of machine may be quite adequate. If we reckon in 4-weekly cycles (21 days plus one week for getting ready for the next hatch) it means that 150 per machine per month will be feasible and so multiples of these machines will be required to meet whatever target is required. If four machines are used, set in different weeks, it should be possible to hatch around 112 eggs per week at 75% efficiency, more if better results can be obtained.

Figure 9.1 A Medium Size Incubator
Bristol Incubator; 700 pheasant eggs (Patrick Pinker Game Farms)
Developed for the hatching of Game bird eggs such as pheasants.

Figure 9.2 A Petersime Incubator
Cabinet type with fully automatic controls.

If a larger throughput is necessary the next possibility is a machine which holds more eggs than the conventional still-air incubator yet does not go up to the size of a cabinet-type machine.* Some of these still have the automatic turning with a mechanism which revolves the trays which are set at an angle. Others are flat machines like the still-air type. In appearance from the outside they look like large refrigerators.

The cabinet machines proper are extremely large and many are large enough to walk in. They circulate the air by means of fans and have very sophisticated controls which allow maximum results to be obtained. These are described below. It should be noted that older type machines may not have the many sophisticated features of the newer models so allowance must be made for this fact.

Medium Size Machines

The medium size machines will suffice for the poultry farmer who wishes to hatch his own eggs. If there is to be a turning back to the pure breeds for free range these machines should find a better market. In the last 25 years or so the large hatcheries have dominated

*References to 'cabinet machines' usually means the large multi-shelf or tray incubator which will take thousands of eggs. Possibly the intermediate size should be called 'small cabinet' machines. Sometimes they are called Table Incubators.

Artificial Incubation

the market and hybrid birds have been the main types of birds available. Whether this will continue depends on the trends in production.

A typical table incubator is shown in the illustration (Fig. 9.1). This is an old model, but the same principles still apply. It will take a considerable number of eggs and generally turns them automatically. The air is circulated by means of fans which come on according to a predetermined timetable, thus bringing in fresh air at a regular rate.

LARGE CABINET INCUBATORS

A few large manufacturers dominate the large incubator market the two main producers being Petersime and Buckeye, but others do exist so a careful search of the market is advisable before a decision is made. However, with all things technical, the wise decision is to look to those who have the technical skills and resources for giving advice and for supplying spare parts if required.

Specifications

The cabinet machine is a large incubator which holds a large number of hen eggs and the similar space equivalent in goose or turkey eggs. Special machines are

Figure 9.3 A Buckeye Incubation System

Figure 9.4 Eggs being wheeled into the Buckeye Incubator

Artificial Incubation

available for *ratites* such as ostrich and emu.

The number of hen eggs starts around 4,000 and extends to about 150,000 – more in some models. The sizes vary, but many are over 2 metres high and over 3 metres wide (front), with a length which can be in the region of 4 metres up to 10 metres, large rooms in themselves. The machines are built into a room, thus allowing the walls to be used as part of the building. The larger models are walk-in rooms into which trolleys are wheeled to be stacked in sequence. Everything possible is done to allow the vast number of eggs to be handled as quickly and safely as possible. Very large capital investments are involved so the hatchery must be run along professional lines.

Types Available

The various types may be classified as follows:

1. Single Stage Incubators (Setters)
In these machines the eggs are all introduced at the same time so there is no mixing of different ages. This means that the ideal conditions can be obtained. In the early stages the heat needs to be relatively high; later on it requires to be reduced, usually with a higher level of humidity. All this can be achieved without difficulty. There is no need to aim for an average set of conditions as in a multi-stage type of machine.

2. Multi-stage Machines (Setters)
In these machines batches of eggs are introduced at different times, say, once per week. As many as eight stages may be present in the single machine, although fewer would be better to manage. This has the advantage of being able to set the required number of eggs available and for which there is a market for the chicks expected. Once full it allows economy of scale and yet gives the necessary flexibility to have a steady throughput of eggs.

In some machines the fact that different ages are together means that a compromise set of conditions have to be the aim. In more modern machines it has been possible to create a variable level of temperature so that new eggs are placed at the rear of the machine where it is running at a higher temperature. The trays are moved forward until they are at the front of the machine before being moved to the hatcher.

These machines may be fitted with fixed racks for loading eggs or with trolley loaders which are filled with eggs before being wheeled into the incubator. These trolleys can be plugged into the electronic control system and turning then takes place by means of an electric actuator.

3. Hatchers
As noted earlier, the hatching process requires a lower temperature (see below) to be generated along with a higher level of humidity. It has now become standard practice to produce the ideal conditions in a special machine called a 'hatcher', which is an incubator which provides all the conditions required for hatching in the last 3 days of the incubation period.

The eggs are transferred into special hatching baskets by moving the trays. Usually this is a standard size (132 eggs) for the machines in question.

The requirements for successful hatching are well known; **a relatively high humidity is essential as well as a very adequate flow of air.**

The view generally taken is that overall temperature should be fairly high when the hatch is occurring, but, if too high, research has indicated that a very high temperature from the machine tends to impose stress on the chicks. They are exerting energy to escape from the shell and therefore do not want too high a temperature, yet this must be high enough to stimulate them to continue to try to emerge from the shell.

Increased air flow is vital and so is the extra humidity or the membranes will dry out and become so tough that the chick cannot free himself. This extra flow of air supplies the oxygen and, at the same time, since ventilation is improved, the excess carbon dioxide, which would be harmful to the chicks in large doses, is taken away. The forced draught machine can ensure that this is done, but the fact that increased ventilation may also reduce humidity, must be allowed for in the calculations when setting the controls of the machine.

4. Special Purpose

Usually the standard incubator takes hen, duck and turkey eggs. The very large eggs such as ostrich or emu can be hatched in modified standard incubators or in special incubators made for that purpose. Since the ostrich egg is very large (1,500 g.) and not usually as plentiful as hen eggs the incubators are designed to accommodate the likely requirements. The Buckeye machine may be for 60 ostrich eggs (12 per tray) or for as many as 2,500 eggs.

Controls

As would be expected, the controls form an integral part of the incubation system. They may be through a **computer which is programmed to monitor each phase.** In addition, or alternatively, **electronic controls** may be employed to watch every stage of incubation.

The various essential stages and processes are dealt with as follows:

1. Ventilation

Paddle fans are employed to move the air being activated by an external motor. There are also adjustment controls.

2. Humidity

Humidity is supplied by having a water tray; in some models there are refinements such as a fine spray on to a tray, or an ultrasonic spray, which gives a very fine spray into the

Artificial Incubation 127

Preparation

Testing (Candling)

Figure 9.5 Systems for Preparing the Eggs & for Testing
Courtesy: Petersime

atmosphere. Compressed air may be used in the process. Control is through a wet-bulb temperature probe set at the required level.

3. Heating & Cooling
The incubators are made of metal panels with thick plastic insulation. This very high level of insulation means that once the prescribed level of heat has been reached the incubators rely on cooling for control, only turning on the heating elements when the temperature falls below a set point.

4. Alarm Systems
The modern machines have sophisticated alarm systems which indicate when any important factor is not working as planned and set in the machine. This can mean failure on heat, ventilation, humidity or any part which is not functioning properly such as a fan or the door not being closed.

CHOICE OF SYSTEM

The system to select will depend on a variety of factors, including the business objectives and availability of finance as discussed in the earlier chapter on this subject. This is a field for the professional breeder and careful assessment is essential before deciding the answers. Testing and related matters are considered in the next chapter.

10

INCUBATOR PROBLEMS

Recognizing the problems

In the process of incubation many problems will be encountered and the breeder must learn to recognize the likely causes. A summary of possible reasons for failure to achieve a reasonable level of hatching is given in the Table overleaf. Unfortunately, it is not possible to be precise or exact on the failures **because generally there is a combination of a number of factors.**

As shown in Chapter 1, there are four main factors and three of them are related to each other. For example, if ventilation is inadequate this will affect the temperature; in turn, this will influence the humidity.

There are **tendencies** which occur **rather that positive laws** so the possibility of some other factor must not be overlooked. We do know that there is a tendency for chicks to be large if humidity and temperature are high (conversely if low), but this can also occur if there is a low temperature and high humidity. Often it is the ventilation that is the main problem.

Where there is a separate hatcher or a single set—

Incubator Failures Table

Problem	Possible Cause
Clear eggs (infertile)	No evidence of being fertile (clear); faulty male or female; inadequate feeding; eggs over 7 days old. Candling rejection.
Late Hatch	Low or variable temperatures; stale eggs; incubator failure.
Early Hatch	High temperature
Chicks Dead in Shell (small air space)	Poor hatchability; inadequate ventilation (machine or room); low humidity.
Dead in Shell (normal air space)	Weak germ indicating stock problems (see 'Clears'). Opening incubator when hatching starts.
Dead in Shell (large air space)	Lacks humidity & excess ventilation.
Hatch wet & sticky (a) shell fragments attaching (b) smeared with contents	(a) Temperature too high; low moisture. (b) Temperature too low; high moisture; poor ventilation.
Dead with Blood Rings	Temperature too high or low; electric failure; eggs gone cold.
Broken Yolks	Rough turning; old eggs; constitutional (heredity); bad handling pre-incub.
Crippled Chicks	Wide variations in temperature.

Figure 10. 1 Table of Possible Incubation Failures

Artificial Incubation 131

ting in the incubator, the conditions can be varied to suit the hatching chicks and this is excellent. However, in the small incubator with different aged eggs, an average must be sought and, what this should be, will only be determined by experience with the particular incubator. **The warning never to add new eggs when chicks are about to hatch or during the hatch must be taken very seriously because this is a common mistake by the amateur breeder.**

If mistakes are to be avoided or their effects reduced to a minimum, it is necessary to make an analysis of what has occurred in each hatch and the likely causes. A basic record is shown in the preceding chapter. The analysis should now be taken further and a feedback system introduced.

Analysing the Results as the Hatch Progresses

Fanciers new to incubation find the checking and repeat checking rather tedious and wonder: Is this really necessary?

In reality, if ideal conditions are achieved (eg; fertile birds, satisfactory accommodation and sound feeding) and the incubator is working as intended by the manufacturer there is no need to test the eggs by candling. The eggs will hatch and there should be no problems.

In practice the testing is vital for a number of reasons:

1. Early in the season the fact that the birds are fertile will be shown by the eggs developing or, if infertile, by clear eggs.
2. The clear eggs may be used for dog food or boiled, chopped up finely and mixed with chick crumbs for the chicks which are from an earlier hatch. Some breeders advocate eating the clears up to 8 days old and no doubt it is safe so to do, but there seems to be something wrong with the notion that they are fit for human consumption. I prefer eggs freshly laid!
3. All eggs, whether fertile or not, use up incubator space and consume heat so it is more efficient to remove the 'duds' and replace them with fresh, fertile eggs.
4. Eggs which are not 'alive' may turn bad and contaminate the air in the incubator. At the worst, if the incubator is not cleared for each batch, a bad egg may stay in for weeks and in the end will start to ooze a foul smelling liquid or may explode; either way the remaining eggs would be adversely affected.

The message is to watch the incubator and monitor the results very carefully. At the end of each hatch all eggs should be removed and if possible once per month the eggs should be transferred to another machine and the incubator washed down and fumigated. If this is not feasible make sure that all new eggs are sanitized and,

Artificial Incubation

when dirty, wash them in washing-up liquid or, if on a large scale, by an egg washing machine and an approved detergent and disinfectant combined. Attention to cleanliness and hygiene generally is absolutely vital for all incubation, but especially so when an incubator is continued through the season, possibly four to six months with some fanciers, shorter with others, and when eggs are constantly being removed and replaced. In the warm atmosphere of the incubator all unwanted germs will multiply quickly if allowed. In selecting an incubator make sure that the parts are easily removable and washable because it may be necessary to wash them down rather speedily if anything goes wrong, such as broken eggs or suspected disease problems. Obviously any live eggs removed must be placed in a warm place and the incubator should be warmed up properly before replacing the eggs.

Remember any changes in the incubator, such as opening it up, washing down, removing chicks, electricity failure or other occurrence will affect the balance previously achieved and therefore this must be corrected before it can be assumed that the eggs are back to normal incubation.

Do not take risks with the eggs which have taken so much effort to produce and in which potentially top class birds await to be hatched.

The Candling Process

The testing of eggs should start in the pre-incubation period and any that have abnormalities externally should be rejected. The **pre-incubation candling** will show whether there are blood or meat spots or other matter which should not be there. It is also wise to check on colour by taking a sample egg from different batches.

We can then assume that all eggs put in the incubator are fresh and clean; the latter is vital for reasons covered earlier. It is also essential to be able to see whether the embryo is developing. A dirty egg is extremely difficult to candle.

The Means

As noted in the chapter that covers the historical development of incubation, various devices have been used to check the eggs. Essentially, the requirement is a **strong light such as a torch or light bulb with a hole the size of an egg so that the light is directed and concentrated in one place, thus seeing the outline of the contents.** In a commercial enterprise it will be necessary to have a light box over which a tray of eggs can be placed and any clears or infertiles can be detected. With thick shells or those which are dark coloured it may be necessary to have ultra-violet light.

Artificial Incubation

Analysis of Results

Date: .... Incubator: ....
Hatch Code: .... No. of Eggs ....

	No.	%	Note on Results / Action Taken
Candling			
7-Day Tests			
Clears			
Broken Yolks			
Addled Eggs			
14-Day Test			
Embryo Died			
Dead in Shell			
Pipped but Died			
Weight Test			**Air Space Check** Measured against Predetermined Ideal
Day 1			
Day 7			
Day 14			
Comments:			

Figure 10.2 Analysis of Incubation Results

In addition, the **Weight-Loss** or **Air Space Tests** may be applied, thus giving a more exact check on the eggs. As noted earlier, an average of 12% is expected to be lost, although in some bird as much as 16% is expected. Parrots apparently have a loss of around this higher figure and this varies because the egg shells differ in porosity from parrots which inhabit different altitudes.* The higher the altitude the lower the porosity to compensate.

The various reasons for rejection of eggs are shown in Figure 8.1 in Chapter 8. They are as follows:
1. **Clears or Infertile Eggs.**
2. **Broken Yolks.**
3. **Dead in Shell.**
4. **Pipped but Failed to Hatch.**
5. **Cripples or Malformed Chicks.**

These are now explained in more detail. **The description is for eggs with a 21-day hatching cycle and other eggs need to be tested at the appropriate times.**

Clear or Infertile Eggs (7/8 day Test)**

The eggs in this are devoid of any germ which will develop into a chick. Accordingly, The egg is quite clear so the blastodisc cannot develop into a blastoderm. A simple test, which can be performed by the fancier, is to hold up the egg, gripping by the first finger and thumb,

See: *Parrots: A Complete Guide* , **Rosemary Low, Merehurst, London, 1988. ** See table of hatching times to calculate when tests should be carried out.

Artificial Incubation 137

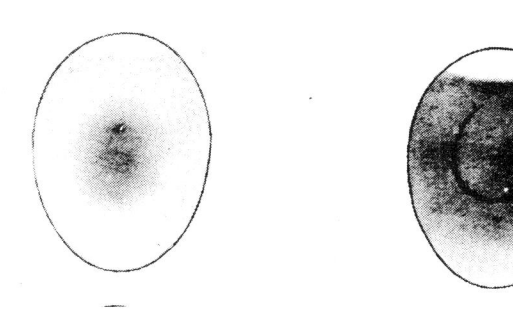

Figure 10.3 The Clear Egg (infertile) & Broken Yolk

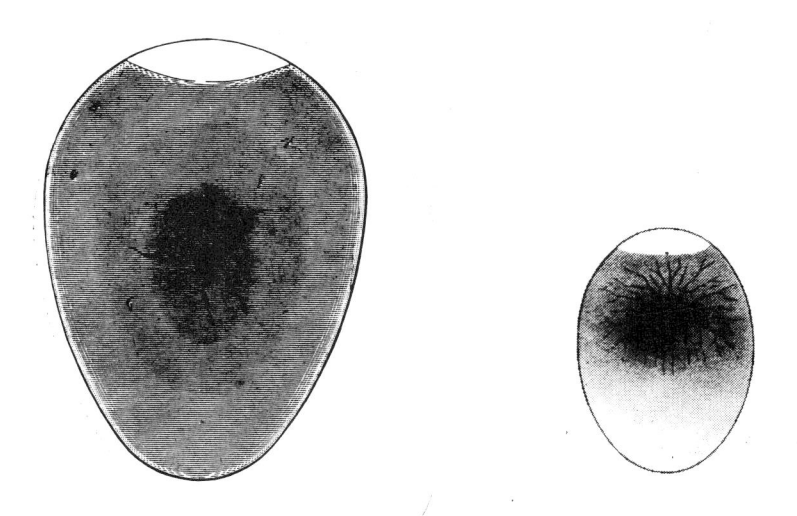

Figure 10.4 The Fertile egg
Note the air space has changed very little.

and look at it with the arm upraised towards the sun. The light shining through will indicate a spidery growth or a quite clear egg, with possibly a yellow tinge around the yolk.

When large numbers of eggs are involved or the days are without sun the candling device must be used. This will be the small light box or torch or for the commercial hatchery a special light box over which trays of eggs can be placed.

As noted, the infertile eggs ('clears') will show no development so the vision is of an empty egg with possibly a slight yellow glow from the yolk and larger air space from new laid eggs.

In the case of **normal, fertile eggs** there will be a dark spot (the embryo) with a network of veins running from it, often referred to as a 'spidery' growth. This should be clearly discernible from the candling. The outline should be positive and not blurred or showing signs of the embryo dying (see below). **These are the possible successes; they now go on to the next stage of incubation.**

The possible outline of the main stages found at the 7 day test are shown in Figures 10.3 and 10.4. This is the most positive test for recognizing the developing embryo. However, do not attempt to test too early on or

Artificial Incubation

some eggs may be rejected even though they are fertile. Also do not leave until the embryos have gone beyond the 8-day stage because then the test becomes more difficult because of embryos which have died although quite well matured.

Broken-Yolk Eggs

These are the eggs which are fertile and should have hatched, but will not progress because the yolk, the mainstay to life and development, has broken. The broken yolk is recognizable by the presence of a 'cloud' in the egg which moves when the egg is turned from side to side. There is a faint line around the yolk and the embryo cannot be seen clearly as in a growing, live situation.

If a broken-yolk egg is shaken gently from side to side the contents may be heard as the liquid moves around in the shell. However, this is not a practice which is recommended to the amateur and this test is best left until after the end of the normal hatch period as a final test. If eggs are left for a few dates beyond the hatching date then give them the test and the result will be obvious. Sometimes this *looseness* within the egg occurs even with broody hens which indicates the condition may be inherited. **In an incubator it is usually attributable to a high operating temperature.**

Dead Embryos

This test is possibly the most difficult because an embryo is present, but is not developing. I always hesitate to discard such eggs because they may turn out to be still alive, but for some reason are not as positive and clear as the ideal suggests. The 'spot' of the embryo is certainly present, but the network of veins is not functioning and therefore it does not show at the candling. If there are many of these eggs the test may have been carried out too early.

Other signs are a muzziness of the embryo and the air space is indistinct. It is no longer alive and therefore the deterioration of the growth has started and is no longer an overall, dark colour. This death may occur at any time, but can be distinguished most clearly if it occurs between 4 and 6 days, although the latter may not be recognizable at a 7 day test – hence the suggestion that it might be better to wait for the second test.

Addled Eggs

The eggs which have gone bad because of bacteria getting into them should be removed at this 7-day stage. Usually it is spotted by the existence of a line within the shell and some discoloration. There may be hair cracks or the eggs may have been exposed to dirty conditions

Artificial Incubation 141

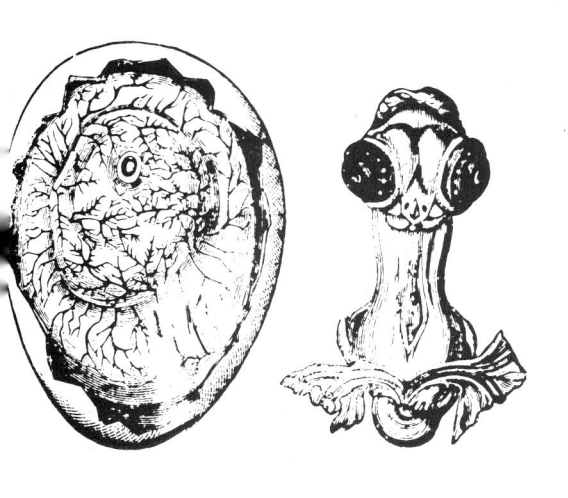

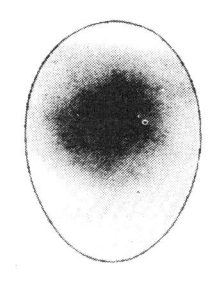

A 5-Day Dead Embryo

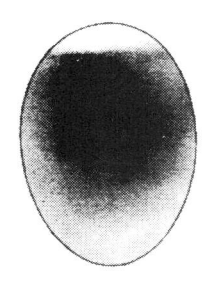

A 14-Day Dead Embryo

Figure 10.5   Dead Embryos

before incubation and not sanitized. Germs are on the shell and enter the egg when incubation starts.

In some cases excessively high temperatures and humidity may cause the eggs to go bad.

Infertile Eggs (14 day Test)

The eggs in the incubator will have been tested once and all clears and defects will have been removed. This later test is concerned with those which were uncertain at 7 days and those which, although alive originally, have perished at the second stage.

The infertile egg will still be clear and after 14 days there is clearly no hope and should be removed.

In the case of embryos which have died the growth expected will not have occurred and instead of being very dark and distinct the egg will be a mottled colour or a dark brown as opposed to a 'black'. The air space will not have shrunk further as expected.

The addled egg will now be a confused mass with the air space which is indistinct and hazy. Compared with the live eggs it will be apparent that there has been no growth of an embryo. Even the 'feel' of the egg will be different from the healthy egg, which tends to start feeling solid with a hollowness to its structure. It will be appreciated that the liquid is being used up and the shell

Artificial Incubation

is going through the process of becoming brittle as it is made ready for the hatching time. Just as the embryo uses up the albumen, and, finally, the yolk, it also draws from the egg shell.

In connection with dead embryos it is worth noting the condition known as a 'stuck germ' which may occur sfter the seventh day. This is where the germ, for what ever reason, has become dormant. There has not been the vigour to go on growing or the egg has become chilled or it is that part of the batch which are simply not capable of being hatched. Whatever the reason, the egg will be in the same state as at the seventh day and therefore must be discarded.

Final Testing (18th to 22nd Day)

At this stage, from 15 days onwards the testing should cease. However, if there is some doubt because of a mishap or it is a fairly easy matter to put a tray of eggs over a light box at the 18th day; when the transfer is being made to the hatcher, then this may be done. More likely, a final test will be made at day 22 when the egg has failed to hatch. Sometimes there are delays due to the age of the eggs or variations in the temperature or other factors. Generally though the eggs should hatch within a day of each other or there is something amiss.

Conditions for Testing

The tests may be made in a simple way as described by looking at each egg before a strong light or even the sun. However, if fairly substantial numbers of eggs are involved and the breeder wishes to be certain, and carry out the work properly, **then there must be provision for darkening the room and thus allowing the maximum light to shine through the eggs.**

A light box with a glass top, rather like those used by graphic artists to plan work, or a metal box containing a strong light and a hole over which the egg can be placed can be purchased or can be made. The examples given in the historical section may be modified and quite easily made; all that is needed is a simple light fitting and on and off switch fitted into one of the devices.

Figure 10. 6 A Light Box for Testing a Tray of Eggs

Artificial Incubation

THE HUMIDITY & AIR SPACE TESTS

The testing at seven days and then 14 days is to check whether embryos exist and whether they are alive. This is a very important part of incubation for the reasons stated earlier in this section. Just as important, probably more so, is the need to **monitor the level of humidity, by checking the air space and, where appropriate, the weight losses at these different stages.**

As noted, overall the main check on whether the moisture is at the correct level is to follow the procedures:

1. Predetermine the air space *standard* (ie, the ideal) at the crucial times:

(a) New laid egg;
(b) 7 Days;
(c) 14 Days;
(d) 21 Days or other hatching date. This will only be used as a check if the hatch has not gone as planned.

2. *Ditto* for the *weight*. In the case of small numbers of eggs for pedigree stock, or rare birds such as parrots, it should be possible to weigh each egg at the different stages. Also a computer with the relevant programme should be able to predetermine the ideal weight loss at each stage, thus giving a standard against which actual results can be compared. This has to be done for each breed or species kept and being incubated.

Where very large numbers of eggs are being hatched the checking can be done on a sampling basis, spotting suitable candidates when the candling is done. A further check is to take the total weight of a tray full of eggs and calculate an average loss per egg for each tray. *Some writers suggest weighing eggs in multiples of 10 which is a compromise approach.*

It means that where checking each egg is difficult the reliance must be placed on the air space. Various records are available for showing the ideal losses (see below).

3. Check and compare the air space and weight loss at each appropriate stage as in 1. above. This will then show whether the humidity level is correct.

4. Where the air space and weight differ to a significant degree the the controls should be adjusted to compensate. However, do not over react or the position will be reversed to cause damage at the other extreme; for example, if the air space is too large, indicating that there is too little moisture, do not adjust too much or the eggs may take in too much moisture and flood the eggs.

The weight loss may be between 12 and 16 per cent and this may occur along the lines suggested in Figure 8.1 on page 109. Thus if 10 eggs weigh 400 grammes the loss expected will be:

(a) 8 days at 3% = 12gm.
(b) 10 days at 5% = 20gm.
(c) 17 days at 10% = 40g.ie, 4gm per egg.

Artificial Incubation 147

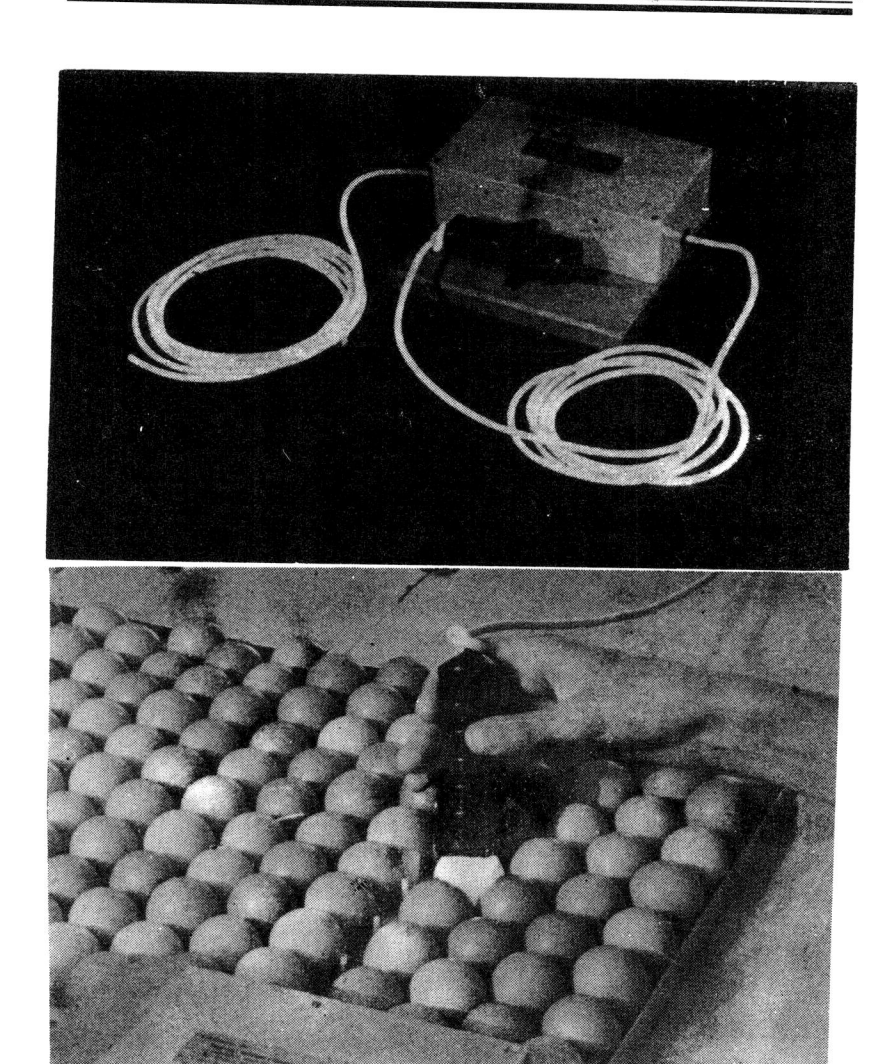

Figure 10.7 Candling Methods.
Above: Small machine. *Bottom:* Testing by hand 'torch'.
Courtesy: Patrick Pinker Game Farms.

Various attempts have been made by poultry scientists to formulate average losses. Thus W A Lippincott found that the depth of the air space **in an egg weighing 2 oz** was:

Day 8 = 23/32in. Day 14 = 26/32in. and Day 19 = 31/32in.

The weight losses (oz) for 100 eggs were:

Day 8 = 13.44. Day 14 = 23.88 and Day 19 = 32.77.

This represents around 16 per cent at 19 days which might be high for some species such as pheasants. The other percentages are 6.72 and 12.94 respectively, all a little on the high side for some types of egg, but clearly successful for some. On the other hand, for geese an expert suggests that 20 per cent should be lost during the incubation period; this seems rather on the high side, but she reports a 70 per cent hatching success rate which is satisfactory for goose eggs*.

Action to Take

If the air space and weights are well away from the standards set for the species concerned then the incubator must be adjusted. In this connection it should be apparent that the 7-day test is of vital importance. Any error corrected at this time should not be

---

Keeping Domestic Geese , Barbara Soanes, Blandford, 1992

Artificial Incubation

too great to rectify the small imbalance which is occurring; if it is very large then it probably means that the initial setting of the incubator was wrong and a serious error will have occurred.

This is not to suggest that an error of, say, 2 per cent too low, should be offset by regulating the machine so that it is 2 per cent too high, thus compensating for the first error and bringing the incubation period back to normal. This does not usually happen, because the normal conditions have been deviated, and, therefore, the only hope is to bring the temperature back to **normal** and then hope the hatch has not been spoilt. The result will probably mean a delay due to the low, earlier temperature.

Fortunately, in modern machines, an error can be corrected quickly and effectively; in forced-draught machines there is little chance of such error taking place because the air is constantly being changed and adjustments are made automatically. With still-air machines, when the regulator has not been adjusted sufficiently, it will be necessary to adjust the ventilation points and moisture tray controls. In some of the older machines to increase humidity, balls of cotton wool, which had been soaked, were placed in the drawers of the machines. Whilst this should be unnecessary in a

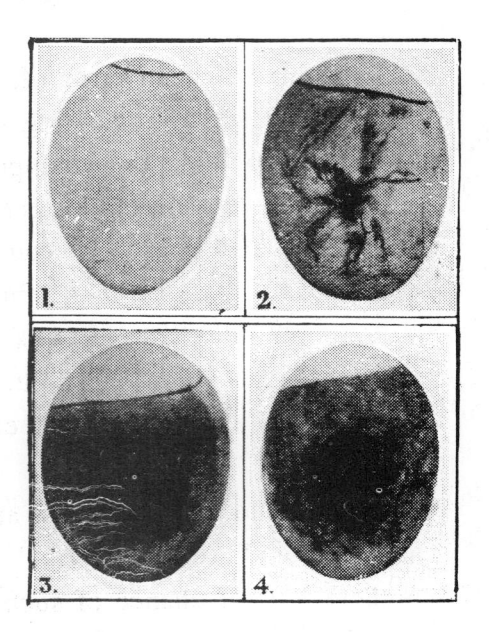

Figure 10.8 Eggs with Standard Air Spaces
Photo: C Grange.
KEY: Photo 1. 1st Day Air Space.
Photo 2. 7th Day Air Space.
Photo 3. 14th Day Air Space.
Photo 4. 14th Day – Too much moisture.

Note: It is sometimes said that as much as a third of the egg will be air space at 19 days and it is surprising how the air space shrinks up to the point where the chick starts to break from the egg.

Artificial Incubation

modern machine it should still be kept in mind if problems of lack of moisture are being experience, when the weather is hot and dry. This could result in embryos drying up and be too weak to hatch.

Watching Progress

The successful breeder relies on his experience and watchfulness as much as the accuracy of the incubator. Inevitably, as the hatch progresses, the egg changes and so does the structure of the contents. For a time the egg seems to be filling out and then, towards hatching time the egg actually feels 'hollow'. With experience the breeder learns the signs and if he or she is to be successful there must be a positive effort to understand the natural progress. In the end it should be possible to handle a egg that is due to be hatched and know that the chick is healthy and likely to hatch. Conversely, where the egg is no longer alive it will have a 'dead' feeling and can be rejected. At first, experiment with eggs which are a few days overdue, when the rest have hatched. Write down what you think is wrong then gently break the shell with a sharp nail sticking from a bench or piece of wood. In this way the learning process can be tested.

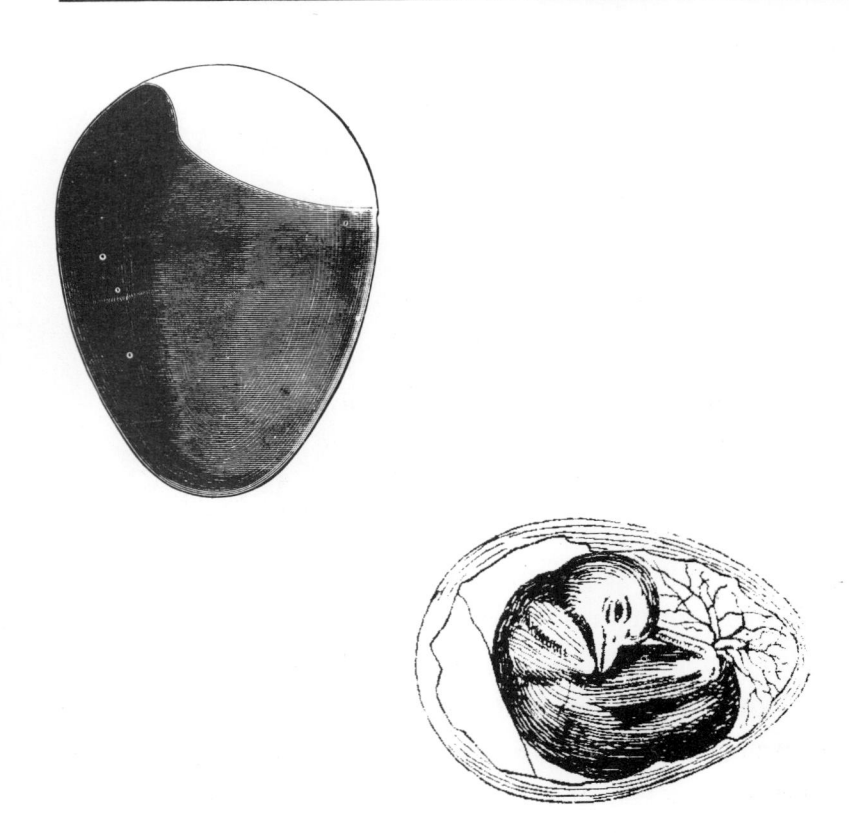

Figure 10.9 Getting Ready for the Hatch

Embryo 3 days before hatching; at this stage the humidity is usually increased with more ventilation, but not excessive or the membrane will dry out and become too tough for the chick to hatch. The proper balance is vital at this crucial stage. See next chapter on the way to achieve the required results.

11
FINAL DEVELOPMENTS & HATCHING

In the preceding chapters the various stages have been examined. This chapter takes the stage to the ultimate goal; that of hatching live chicks which can be reared.

This stage is in many ways the most critical because if conditions are not correct all the previous work will be lost; therefore it is of vital importance to pay special attention to humidity and ventilation.

The Hatching Process

As noted, three days before eggs are due to hatch it is usual to increase humidity and ventilation and, at the same time to keep a high level of overall temperature, but not excessively high or the chick will not be able to give enough energy to emerge from the egg. **The hatching conditions** may be achieved in an incubator proper or in a separate hatcher into which the eggs are transferred on the due date.

The various stages are shown diagrammatically and also explained by giving appropriate notes. The length of time each stage takes will vary with the species concerned so the notes here are intended as a general guide.

The Hatching Date

The date due will have been noted on the **Incubation Record** and three days before the date it is usual to stop turning.

Typical periods are as follows, but remember fresh eggs and those which have been incubated at a fairly high temperature may hatch a day early. It is undesirable to try to hatch earlier than within the normal period.

Breed/Species	Days
Fowl – Large	21
Bantam Fowl	19 to 21
Pheasants– Wild	24
Pheasants – Ornamental on species).	21 to 28 (depending
Partridge	23
Domestic Ducks	28
Domestic Geese	28 to 35
Muscovy Duck	35
Guinea Fowl	28
Turkey	28
Peafowl	28
Jungle Fowl	19 to 21
Ostriches	40
Emus	57
Rhea	35
Waterfowl	Great variation between species; eg Swans 30 to 37; Geese 22 to 35; Ducks 22 to 30.
Quail	21 to 23

Artificial Incubation

Those breeders wishing to obtain specific dates on the more unusual birds should consult a specialized book dealing with the species. Parrots seem to have incubation periods which generally (though not always) correlate with size; from 17 days for smaller species and around 30 days for the larger species, such as Macaws. **Remember though that there may be variations from the normal times, due to the incubator or the eggs.**

STAGES IN HATCHING

As the hatching date approaches the chick completes its final stages as described earlier. The main, final steps are to absorb the yolk through the navel and to push through into the air space so that the work on the chipping out can start. At this stage it will be possible to hear a chick as it moves and sometimes a chirping sound is heard. After a short time, although it may be as long as 24 hours, the chipping of the egg will start.

The Egg Tooth and Pipping

During the formation of the chick a small 'tooth' appears on the top of the beak and this, known as an 'Egg Tooth' is used for forming the grove around the egg for the chick to escape. This tooth disappears after hatching and is of no further use to the chick. The membrane and shell from within the egg are more vulnerable than the outside so the tiny chick by persistent tapping is able to carry out his task over a period as a normal routine. How the chick knows its duty is to emerge from the egg is one of the great mysteries of nature, but the instinct is there and, provided the conditions are correct, the chick will do its work.

Once a tiny outward eruption occurs ('pipping') the hatch has really begun and on no account should it be interfered with.

Note: An exception is where the hatch is being done in a still-air incubator and the pipping occurs on the underside of the egg when the heat is only from the top, then it is advisable to turn the egg to the top. In a large forced-draught incubator the eggs are stacked in trays and it is usual to place the eggs with the broad end uppermost so there should be no problem with warm air circulating or with subsequent hatching; if the narrow end is uppermost the embryo may grow in an abnormal position and have difficulty in pushing through into the air space.

The pipping has now commenced and the chick jerks its head and gradually, using the egg tooth and turning in the egg, makes the channel in the shell.

Artificial Incubation 157

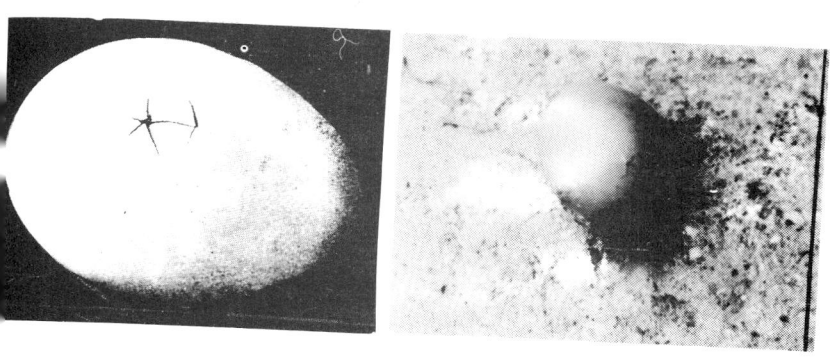

The First Pipping (Hen Egg & Pheasant)

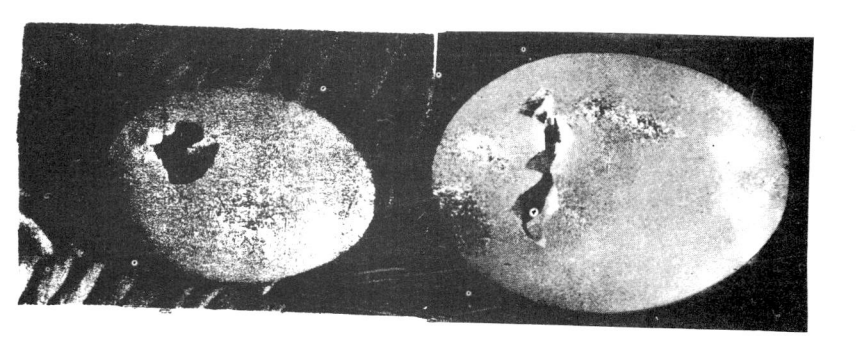

The 'Hole' Appears

An obvious hole is not always present and in some ways it is better not to have too large a hole because this may dry up the membrane (see Fig 11.2). The right hand egg is more usual and preferred.

Figure 11.1 Stages in Hatching

Pecking its Way Out

The chick has started to peck all round the shell and it will now be breathing the fresh air of the incubator. This is why the temperature may go up at this stage (the chicks are working and generating heat) so it is vital to ensure that there is adequate ventilation and the **appropriate amount of humidity** (see below on different species). As noted above, the chick now makes a groove around the egg. This may take many hours, from 6 to 10 being typical. The hole should not be enlarged or interfered with or trouble will be caused because the contents will dry out too rapidly. In any event, the whole process is quite complex, one stage relying on the other, so any interference will upset the balance.

The next step is for the chick to pull itself free from the lower shell. It will do this by grasping the edge of the shell and pulling itself free. It will be exhausted from its efforts and its down will be damp (not excessively so or 'mushy' or having a 'clubbed down' appearance – all signs of problems in hatching). Although it will now rest before attempting to move, its appearance should be one of normality with no signs of breathing or other difficulties. At this stage it may be left in the incubator to dry which will take a few hours or moved to drying trays where the temperature will still be high, but

Artificial Incubation 159

The Top is Free
At this stage the chick puts its feet around the edge of the lower part of the shell and pulls itself out (see below).

Exhausted but Out
Initially the chick will lie flat with its head forward, wet and exhausted and it must be allowed to dry out.

Figure 11.2 Stages in Hatching (cont).

not as high as in the hatcher. There is a dilemma – moving immediately will probably be better, but this will certainly reduce the temperature of the incubator, and if left the hatched chicks may get in the way of those hatching.

Many poultry experts believe that removal every 6 hours or so is advisable because, if left too long, the chicks will be distressed with breathing problems. Hatching boxes may be used as a temporary expedient, but these should be pre-heated to 37.7deg.C.(incubation level) and dropped to 35 deg. It is probably more desirable to transfer the chicks to brooders, also pre-heated to the same level, thus getting the chicks ready for their new quarters. Tests on whether the heat is at the correct level will be apparent from the chicks; if too hot they will gasp for breath and if too cold will huddle together; however, test thermometers will indicate the temperature being achieved and these should be placed awhile in different positions with the lamp 5cm above floor level.

Once all the chicks have hatched they are placed in the brooder to dry and recuperate from the hatching. **Give no food or water for 24 hours and after that do not expect them to eat or drink much, but if provision is made it only needs one or two to start pecking and others will follow.** The early sustenance comes from the absorbed yolk.

Artificial Incubation 161

The Hatched Chicks now Dry

Chicks in Hover
Watch to make sure chicks are not exposed to cold air at the early stages

Figure 11.3 Ready to Start Being Active

SPECIAL CONDITIONS
Pedigree Hatching

If special groups of birds are to be hatched this may be done by using small bags made of muslin or by placing the eggs in a small wire cage which is inserted in the incubator. This may be done when many batches are hatched in a large incubator. The insertion into the special cage or bags is done on the eighteenth day and a label giving full details is placed on the recepticle. The small fancier may manage with simply marking the eggs, although this presents problems because once chicks are hatched they mix together.

Once hatched the special batches should be ringed with special colour rings or closed rings may be used. At this stage only small rings will be possible for large rings will catch or fall off. Alternative methods are to place wing bands on one of the wings (aluminium, numbered tags secured by rivets or attaching to the elbow by slitting) or by toe punching; this consists of punching a number of holes as a code to the batch. Dye may also be used and with valuable birds a tatoo method may be employed. The details of whichever method is used will be recorded so that each batch can be identified.

This method should be used in conjunction with **Trap Nests**, whereby the hens are kept in the nest to record exactly the producer of each egg.

Artificial Incubation 163

SPECIAL HUMIDITY REQUIREMENTS

The humidity and related requirements have been given in various chapters, but it should be noted that some breeds or species require special attention. Much depends on the incubator and other factors, but the following should be noted:

1. Ducks and Geese
These require extra moisture so the incubator is run as normal at 55 to 60% RH, but for the last 3 days increased to 65% (32° C. wet bulb). Goose eggs in particular must have a lot of moisture and some spray the eggs with warm water.

2. Quail
High at first (58% RH) for 12 days then 55% for 16 days, increasing to 75 to 90% for the hatching which is usually quite quick, then reduce to 53% for the chicks to dry.

3. Guinea Fowl
Normal at first then to 75% at 3 days before hatching.

4. Turkeys
Should be run at 60% normally then at 75% from day 24.

5. Pheasants and Partridge (including Peafowl)
Not too much moisture at first (55%) then increase to 70% at 21 days.

Relating to Temperature Readings

These RH percentages are only a guide; much depends on the type of machine. For guidance on this subject it should be appreciated that if the temperature is 99.75° F (37.6 deg. C.) then the Wet Bulb should be in the region of 90 to 94 deg. F. for the last 3 days, with those requiring extra moisture on the higher level, as indicated above. The *normal* wet bulb reading (ie for main part of the hatch) should be in the range 83 to 88 deg. F.; those requiring the higher levels are pheasant, ducks, and geese. Once the hatch is completed the humidity should go back to normal.

Note on Relative Humidity

Relative Humidity is of vital importance. Moreover, it is affected by the operating temperature of the incubator. Basically it is the amount of moisture in the air and when the humidity is increased the temperature should be reduced. Otherwise, there is an imbalance in the factors covered in Chapter 1.

An extension of this basic definition is:

The water vapour contained in the air compared with the amount that could be carried at that temperature. This may be expressed by reference to the temperature on the Wet Bulb or as a percentage, which is more meaningful.

As noted earlier it is necessary to read the figures from two thermometers:

1. Dry Bulb (normal thermometer)
2. Wet Bulb (surrounded by material which is moist)

When each reads the same there is 100% RH and only when the two differ will it be possible to see when the desired humidity (55 – 60% generally) has been achieved. The wet bulb reading can be compared with a chart which shows the RH percentages and in this way the correct percentage is achieved. However, following a predetermined **Wet Bulb reading** (for the incubation temperature) the same result is achieved. This is explained in the preceding section (page 163).

12
REARING

Historical Development

The first attempts at battery brooding started around 1910 and the initial attempts failed because chicks developed leg weakness. It was subsequently discovered that the main cause was lack of the Vitamin D which is supplied by sunshine. There are other types of leg weakness which cause birds to sit down and rest, but this appears to be a different problem, possibly associated with growth which is too rapid. Once the nutritional problems had been solved considerable progress could be made.

The fact remains that great care must be taken with the feeding and conditions, including housing, if artificial rearing is to be a success. Moreover, these days much more attention must be paid to the health aspects to ensure that birds are not overcrowded or for that matter subjected to any type of suffering.

Chicks require three essentials:
1. Warmth at the correct level
2. Good quality food given regularly
3. Appropriate accommodation, free of dampness and draughts.

There are many ways of rearing young stock. Usually, for birds which are to be bred from, it is desirable to rear them with access to the outside world of grass and sunshine. With poultry stock this movement to an outside environment can take place from about 4 weeks of age because then they can be on a lower temperature.

The methods available range from what is a heated box (for a fancier) to huge, factory-like buildings where the environment is entirely controlled, including food and water, and broilers and other birds are brought to the required levels of maturity within clearly defined periods. A possible classification is as follows:

1. Infra-red lamps suspended in an enclosed space which can be enlarged as the chicks grow.

2. Hovers which are special box-type devices for heating up the air and providing the correct rate of ventilation. As the chicks grow they are allowed more space and the heat is reduced.

3. Brooders

These consist of tiers of compartments or cages in which heat is provided at different levels. Each week or other period the chicks are moved on to a cooler section or the heat is turned down in the existing compartment. Food and water can be supplied in special hoppers or even automatically.

4. Intensive Rearing on Deep Litter

The basis is a room in which heat, light and ventilation is provided. The aim is to provide the ideal conditions for rapid growth so that the birds can be marketed at the best possible time for maximum food conversion.

5. Outdoor sheds and hay brooders

Selection

The choice is very much influenced by the scale of operations. If a few chicks are to be hatched and reared it might be better to use a broody hen or a combination of hen and small incubator.* This is the most economic approach, the main handicap being the availability of suitable broodies at the appropriate time. If more than 35 chicks are to be hatched and reared then an incubator may be worthwhile, but for a very small number the effort and cost may not be justifiable.

The choice for the fancier who rears a few birds is as follows:

1. Infra-red Lamp
2. Hover
3. Brooder
   (a) Under floor heating
   (b) Lamp or heating element

Each has its own advantages or disadvantages, mainly on cost factors. Examples are shown overleaf.

* See *Natural Incubation & Rearing*, Batty J, where full details are given.

Figure 12.1 Infra-red Lamp

The lamp is set at 45cm from the floor and raised 5cm each week, thus regulating the temperature. More space can also be given so the chicks become quite hardy.

Figure 12.2 A Typical Hover

These were very popular and are ideal for the small poultry keeper where 50 or more chicks are to be reared.

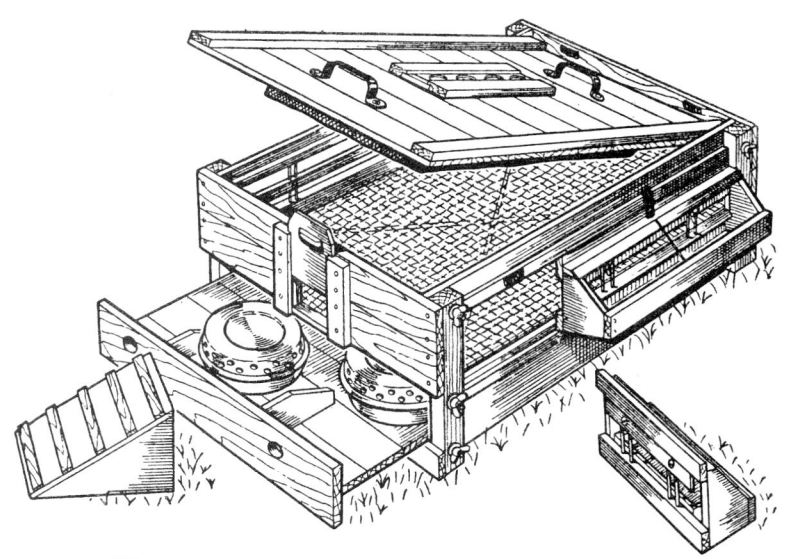

Figure 12.3 A Warm Floor Brooder
Heat can be supplied by lamps under the false bottom.

Figure 12.4 Floor Brooder for use in a Room/Shed
The heating element is inside the floor and warms up. Other systems use a 'mat' rather like an electric blanket. Make sure there is no danger of wires being exposed.

Notes for the Small Breeder

The small breeder can manage quite well with an infra-red lamp of 25 watts. This should be adequate for 50 chicks at a time. It is set about 45cm from the floor and raised each week until the temperature falls from around 35 deg.C. to the point where they are no longer in need of tne lamp. Thus at 5 weeks the heat will be on around 21 deg. C. and only warmth should be needed in the evenings. In the late spring or summer, once the chicks are feathered there is no need for any heat. In fact, some form of hay brooder may be supplied so the chicks can get sufficient heat from the 'packing' or lining, plus their own heat.

With the lamp method it is necessary to keep the chicks within the heat arc of the lamp and therefore some form of hardboard surround will be advisable or the lamp should be at the end of a small shed, well insulated, with the walls forming a natural barrier. As the chicks grow more space is given by extending the run.

Floor cover is vital. New shavings are ideal, but not sawdust because the chicks will try to eat it. Some breeders have an earth floor and give the chicks turves dug from the garden in which they scratch, which gives excellent results because the chicks are kept active. It also exposes the chicks to outdoor materials and provided the soil is free from disease gives them 'natural conditions'.

Transfer to a shed which receives sunshine, even if only through wire netting will give much better feathering than totally-controlled intensive rearing; outside runs from about 5 weeks are vital, if feathering is to be up to show standards. Intensive rearing beyond 4/5 weeks cannot really be recommended for the fancier or those who are producing breeding stock, especially if they are later to be on free range.

The first few feeds can be on cardboard or on papier mache egg trays, the **chick or turkey crumbs** being sprinkled on the board or tray and the chicks encouraged to eat. After that a **shallow metal or plastic tray** should be used and then after a week or so use small **hoppers** which allow chicks to feed without scattering the food by scratching. There should be adequate trough space for all the chicks to feed. The turkey crumbs are higher in protein and therefore bring on chicks quicker, but the chick crumbs, with around 18% protein, have a drug which gives some protection against **coccidiosis**, a curse amongst some breeds of poultry.

A **water fountain** should be filled daily or more often as required and the chicks should not be allowed to be without water. A shallow-troughed fountain is essential, thus avoiding the danger of drowning.

Larger Small-Scale (50 Chicks plus)

Where more than 50 chicks are to be reared one of the alternative systems may be used such as the hover, brooder or the heated floor brooder. The same general observations apply. If electricity is difficult to use because of location there are gas or paraffin appliances.

Moving Outside

Once the chicks are feathered they can be put outside in a run. If a wire-netting run with a movable shed is to be used then make sure no predators can tunnel underneath (put netting on the base of the run). Growers' pellets and mixed corn should have been introduced to the chicks at about 4 weeks of age; in fact, broken corn, including maize, should be given at around 3 weeks of age, along with chick weed and other greens, such as grass clippings, thus gradually breaking away from a diet of only chick crumbs. The latter should be discontinued at 5 or 6 weeks because by then the chicks will be accustomed to the more normal diet.

The Sussex Ark and similar small sheds can be used to bring on the stock to maturity. At about 4 months they can be separated into the future breeding pens, although it may be wiser to introduce the cock bird when fertile eggs are required. Provided he is quite fit and has been in an outside run eggs will be fertile in about 2 weeks.

Artificial Incubation

Figure 12.5 Intensive Rearing
Gas brooders are used and food is automatic as well as water.
(Courtesy: Maywick)

LARGE SCALE REARING

This is the concern of the specialized rearer who may be involved in rearing many thousands of birds for his own use or under contract for layers or broilers. The most popular method is **intensive rearing in a large building** where every aspect is controlled – temperature, feeding, water, lighting, ventilation and so on.

Great attention is paid to hygiene and the birds are vaccinated against possible infection.

The heating may be provided by hot water pipes or by heaters operated by gas or oil. Fans may also be used within the system to circulate the air. The correct temperature, as discussed earlier, is vital to growth and the maximum utilization of the food given.

Water and food is fully automatic so there is no danger of the birds being deprived. However, in these circumstances very careful records have to be kept of food and water consumption and there must be regular checks on the system to ensure that no faults occur. Any breakdown can be very costly if not spotted within a very short period.

Deep litter of wood shavings should be spread over the floor to a depth of 15cm and this provides comfort and warmth.

This system calls for skilful management and proper training is therefore vital.

Lighting

In intensive systems maximum growth can be achieved by providing lights on a controlled basis. Maturity can be controlled by the lighting pattern used. Giving maximum light (24 hours) at the beginning results in maximum growth, but this is then slowed down

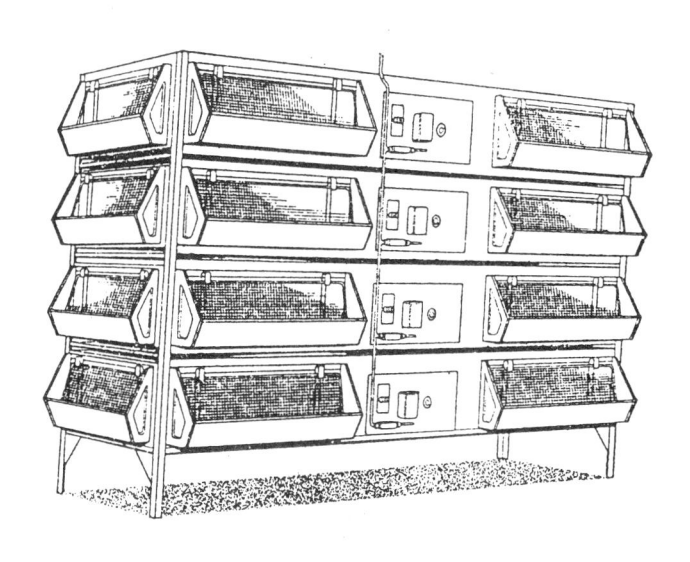

Figure 12.6 Battery Brooders
These are suitable for rearing young chicks

to delay laying; this is known as the **step down system**. A pattern of about 12 hours per day may be adequate, but much depends on the purpose of the birds. *Only a dim light is necessary and advisable.*

NOTES ON SPECIAL REQUIREMENTS

An efficient incubator should be able to hatch any type of eggs and therefore in theory one will do just as well as another. In practice, it has been found that certain types of incubators give better results.

1. Size

An incubator is made for a specific number of eggs and these may be placed in the capacity available as instructed by the maker. Where provision is made for different ages to be included this is acceptable, but only within the recommendations of the manufacturers. With the small incubators at least 50% capacity should be filled or top results may not be obtained. Running a 50-egg incubator with 6 eggs is uneconomic and is not likely to be efficient.

2. Still-Air Or Fan (Forced Draught)?

Both these have their merits and there is a tendency to assume that the still-air is more appropriate for a small number of eggs or the forced-draught is only for large numbers.

In practice this is not the case. Exotic birds such as parrots are best hatched in a small forced-draught incubator. It is said to give greater stability in the level of heat and humidity.

On the other side, those concerned with specialized birds such as ducks, geese, pheasants and partridge have found results better with a still-air incubator.

REARING DIFFICULTIES

In artificial rearing difficulties have been experienced in getting the correct conditions for some types of birds.

Game Birds

Game pheasants tend to be reared in outside brooders, thus giving them early access to outdoor conditions. Special brooders have been developed for this purpose which withstand harsh weather conditions.

The feeding and watering is also organized so that the task is made as simple as possible with appropriate feeders and drinkers.

Ducks & Geese

These can be reared like chicks, but they tend to be rather messy. In any case the ducklings or goslings should be placed on the range so they (the goslings) can graze as soon as possible. If the weather is inclement cover the run with plastic or polythene sheeting so it is rain proof. Feed turkey crumbs initially and then growers' pellets and corn to supplement the grass. Take special care with the water facilities so the area does not become water logged. A special fount or a fount on a twilweld stand will allow excess water to drain away.

Commercial Quail

These present no difficulty in rearing, but they are so small and independent that they need special watching. The wire for the cages used needs to be a very fine, aviary-type mesh or they will escape.

Turkey crumbs are used for feeding and chick founts for water. They are fed on these for ever and they start laying at 12 weeks of age. A lighting pattern is essential for rearing and laying – around 17 hours for laying.

Ostriches

A high quality food must be fed and ostrich chicks should be kept indoors for about 8 weeks and then let out when the weather is dry. A steady growth should be the aim with special rations which should include a protein level of around 15 per cent.

An infra-red lamp can be used for the initial rearing just as for chicks, but allowance must be made for the fact that birds grow very rapidly.

Figure 12.6 Outside Brooder for Game Birds
For full details on the rearing of pheasants and partridge on a large scale see Book List p180.

Artificial Incubation

Bantams

These miniature poultry are quite hardy and after the first 2 weeks are quite strong and healthy. Take care that the temperature is maintained steady and they are feeding and drinking in the early stages.

Parrots and Exotic Birds

These have to be fed by hand a number of times each day so the operation is very time consuming as well as requiring great patience. 'Brooders' have to be provided, usually a small plastic box within an environment which starts at a temperature of 98.5deg.F. (incubation temperature) and is reduced by one degree after drying and reduced to 95 and then 93, counting in weekly periods. Paper napkins are used as bedding. Once feathered the temperature can be reduced to around 80deg. F.

The food is liquified so that it is taken and digested easily. The ingredients must provide the necessary protein for growth as well as vitamins. It may include dog biscuits and baby food such as creamed-corn and vegetables, all thoroughly microwaved and emulsified. Vitamins are added. It is held and fed by a special spoon or syringe or eye dropper every 2 hours (to fill the crop) and less frequently as the parrot grows.

This is an exacting occupation with valuable livestock so hit-and-miss methods will not be acceptable.

EXTRA READING

This book has covered the essentials of incubation and rearing. For those wishing to pursue some of the more specialized areas readers are advised to consult the following works:

Parrot Incubation Procedures, Rick Jordon, Silvio Mattacchione, Pickering, Ontario, Canada.
Covers all aspects of Parrot incubation and hatching. It is written by an expert in this relatively new field and is essential reading for breeders of Parrots.

Ostrich Farming, J Batty, BPH.
A concise yet comprehensive coverage of this latest development in the live stock farming field.

Keeping Guinea Fowl, John Butler, BPH.
Covers the history and development of domesticated Guinea Fowl with special reference to free range.

Keeping Peafowl, James Blake, BPH.
These exotic birds are now very popular and this concise guide shows how they can be managed.

Keeping Jungle Fowl, J Batty, BPH.
The ancestors of modern poultry; their management explained.

Practical Poultry Keeping, Dr J Batty, BPH.
Essential guide to the techniques and methods available for managing all types of poultry.

Modern Guide to Professional Gamekeeping, J Mason, BPH
Gives detail on methods used for incubating and rearing Game birds.

Parrots: Hand Feeding & Nursery Management,
Howard Voren & Rick Jordan, Silvio Mattacchione.
A detailed account of rearing young parrots of various types. Covers the food requirements and methods employed with many coloured illustrations of birds.

INDEX

A
Addled Eggs 140
Air Space 24, 25, 145, 146, 148, 150
Alarm Sytems 128
Albumen 25, 31
Amnionic Fluid 17
Artificial Incubation 2
    Advantages 3
Avian Egg, Romanoff & Romanoff 34

B
Bird World, Farnham 66
Blastoderm (Blastodisc) 25, 36, 42
Breeding 35
Brinsea Incubators 111
Brooders 166
Broody Hen 1
Buckeye Incubators 60, 122
Building 77 (see Hatchery)
Bulb - Wet & Dry 11
(also see Humidity)

C
Candling 64
Candling Devices 48, 127, 147
Capital Expenditure 2, 67, 71
Capsule 51, 63
Chalazae 25, 34, 42
Champion Incubator 54
Check List (Incubators) 69
Checking Incubator 96
Clears 136
Cooling Eggs 103

D
Dead Embryos 140, 141

E
EGG
  Creation 29
  Formation 29 - 34
  Incubation 17 - 28
  Parts of 24, 25
  Sizes 21
  Suitability 20
  'Tooth' 155, 156
Egypt 44, 49
Embryo Development 37, 42

F
Failures in Incubator 130
Feedback 50
Fertile Egg 137
Floor Covering 170
Forced-Draught Incubator 117
Formaldehyde 81

G
Germinal Disc (see Blastoderm)
Getting Ready 152
Goose Eggs 115
Grit 20

H
Hatching Eggs 22, 23, 125, 158, 159
Hatching Process 153
Hatching Records 14, 15
Hatching Room 2, 86
Hatcher 114, 153
Hatchery 76 – 94
Hearson Charles
(Pioneer on incubation) 51 – 55
Heat 4
History 43 – 66
Hover 161
Hygrometer 12
Humidity 9, 10, 85, 88, 145, 163, 164

I
Incubation 1, 4, 95
Incubation Periods 154
Incubator Controls 61, 104 – 109
Incubators 51 – 61
Incubators – Cabinet 121
Infra-red Lamp 166
Infundibulum 31 – 34
Insemination 35
Intensive Rearing 166
Isthmus 31

K
Key Factors in Incubation 5

L
Lamps for Candling 48, 84
Large Scale 117
Lighting 145

Loading Trays 100 – 102
Low Rosemary 66

M
Machine Operation 98
Magnum 31, 34
Mallee Bird 1
Marking Eggs 98
Marsh A 62, 111

N
Natural Incubation 1
Natural Incubation & Rearing,
Batty J, 1

O
Oocytes 30
Ostrich Eggs 115
Ostrich Farming 52, 123
Ovary 30, 31
Ovens in Egypt 44, 45
Oviduct 30 – 34

P
Patrick Pinker Game
  Farms 119
Parrot Eggs 66, 145, 155
Pedigree Hatching 162
Petersime Incubators 60
Pheasant Eggs 157, 159
Pipping 155, 157
Present Values (Investment) 74
Problems 129

Artificial Incubation

Q
Quail & Eggs 115, 177

R
Ratites (Ostriches, etc) 123, 178
Rearing 165 – 175, 176 – 180
Reaumur de M. 48
Regulator 50, 51
Relative Humidity 11
Reproduction 35

S
Selection of Incubator 68, 69, 99, 110, 175, 176
Setters 84
Shell 18, 24, 31
Spermatozoa 34, 35
Spring Cleaning 97
Stacking Trays 156
Still-Air Incubators 118

T
Temperatures 45, 85, 96
Testing 65, 91, 97, 136
Thermometer 97
Training 2
Trays 85, 93
Turning 1, 13, 103

U
Uterus 31

V
Ventilation 7, 85, 87
Visual Incubators 112, 113

W
Washing Machine 28, 84
Weight Losses 146

Y
Yolk 17, 24, 30, 32, 42

SOURCES OF SUPPLY

Many specialized suppliers serve the needs of the poultry fancier and cage bird enthusiast. Besides various clubs and societies there are:

Southern Aviaries, Tinkers Lane, Hadlow Down, Uckfield, East Sussex, TN22 4EU
Supply a very wide range of products such as cages, feeders, books and other essentials.

Bird World, Alice Holt, Farnham, Surrey.
Experts in breeding and rearing exotic birds. A visit is recommended.

Advertisement

Poultry Equipment by Mail Order

Of interest to the poultry enthusiast and bird fancier is Noah's Ark Products, a specialist equipment marketing company based in Essex and owned by Mrs R Bowerbank.

Delivery to any part of the U.K. within a few days of placing an order is virtually guaranteed. Larger items such as portable pens, coops and broody boxes will be delivered within ten days. Noah's Ark are approved stockists of incubators, brooders, electric fencing, housing, health & hygiene products and a wide range of ancilary equipment.

Electric Nets
Protect and contain your birds.

Books ▪ Books ▪ Books
For the beginner and fancier or smallholder.

Chickens, Ducks, Geese, Bantams, Pheasants, Partridges, Guinea Fowl, Turkeys, Quail, Pigeons, Falcons, Goats, Donkeys, Rabbits, Bats, Vermin Control, Housing.

NOAH'S ARK PRODUCTS

Incubators

NOAH'S FREE OFFER

Poultry Housing
Quality Tanalised Housing

Flat Packed, Easy Construction

NATIONWIDE DELIVERY ORDERS DESPATCHED SAME DAY

MAIL ORDER SPECIALISTS

NOAH'S ARK PRODUCTS (BHP),
1 HIGH ROAD, LOUGHTON,
ESSEX IG10 4JJ

FREE CATALOGUE
Please Telephone or write
Tel: 081-502 3800 Fax: 081-508 2051